P9-ASL-609

Technological Change in the German Democratic Republic

Westview Replica Editions

The concept of Westview Replica Editions is a response to the continuing crisis in academic and informational publishing. Library budgets for books have been severely curtailed. Ever larger portions of general library budgets are being diverted from the purchase of books and used for data banks, computers, micromedia, and other methods of information retrieval. Interlibrary loan structures further reduce the edition sizes required to satisfy the needs of the scholarly community. Economic pressures on the university presses and the few private scholarly publishing companies have severely limited the capacity of the industry to properly serve the academic and research communities. As a result, many manuscripts dealing with important subjects, often representing the highest level of scholarship, are no longer economically viable publishing projects--or, if accepted for publication, are typically subject to lead times ranging from one to three years.

Westview Replica Editions are our practical solution to the problem. We accept a manuscript in camera-ready form, typed according to our specifications, and move it immediately into the production process. As always, the selection criteria include the importance of the subject, the work's contribution to scholarship, and its insight, originality of thought, and excellence of exposition. The responsibility for editing and proofreading lies with the author or sponsoring institution. We prepare chapter headings and display pages, file for copyright, and obtain Library of Congress Cataloging in Publication Data. A detailed manual contains simple instructions for preparing the final typescript, and our editorial staff is always available to answer questions.

The end result is a book printed on acid-free paper and bound in sturdy library-quality soft covers. We manufacture these books ourselves using equipment that does not require a lengthy make-ready process and that allows us to publish first editions of 300 to 600 copies and to reprint even smaller quantities as needed. Thus, we can produce Replica Editions quickly and can keep even very specialized books in print as long as there is a demand for them.

About the Book and Author

Technological Change in the German Democratic Republic

Raymond Bentley

Focusing on East Germany's capacity to innovate and diffuse technology, this book sheds light on the technological gap that has developed between the two Germanies. Dr. Bentley compares the sophistication of GDR and FRG technology in different industrial branches, evaluates the strengths and weaknesses of the GDR's research and development system, compares the R&D effort of the two Germanies, and discusses the government policies that affected technological change in GDR industry from 1945 to 1975. He identifies and analyzes hindrances to research, innovation, and diffusion in the fields of planning, organization, economic stimulation, and ideology, and looks at the formation of interest groups. He also compares evidence from the GDR with data from other countries, including the USSR.

Dr. Raymond Bentley wrote this study as a guest researcher of the Abteilung fuer Wissenschaftsforschung (Department of Science Studies) at the University of Ulm, Federal Republic of Germany.

Technological Change in the German Democratic Republic

Raymond Bentley

Westview Press / Boulder and London

The paper used in this publication meets the minimum requirements of the American National Standard for Permanence of Paper for Printed Library Materials Z30.48-1984.

A Westview Replica Edition

Published in 1984 in the United States of America by
Westview Press, Inc.
5500 Central Avenue
Boulder, Colorado 80301
Frederick A. Praeger, Publisher

Library of Congress Cataloging in Publication Data
Bentley, Raymond.
Technological change in the German Democratic Republic.
(A Westview replica edition)
Includes bibliographical references.
1. Technology and state--Germany (East) 2. Research, Industrial--Germany (East) I. Title.
T26.G42B46 1984 338.9431 84-2262
ISBN 0-86531-812-3

Printed and bound in the United States of America

10 9 8 7 6 5 4 3 2

Contents

Tables

Foreword

Since the 1950s interest has grown in the science-technology systems of Eastern Europe. Especially, of course, the R and D effort of the Soviet Union has been the subject of great curiosity and speculation, and of some serious investigation. The Birmingham University Centre for Russian and East European Studies has been responsible for several outstandingly good studies of the Soviet R and D and innovation systems. Yet very little has been published on the science-technology systems of the other East European countries apart from propagandist tracts and official statistics which are extremely difficult to interpret.

The divergence of West and East European statistical systems affects many aspects of economic analysis, since it includes such fundamental concepts as GNP. International organisations have made praiseworthy efforts to reconcile the differing international standards and to produce some comparable statistical series. However, in the case of R and D statistics and other related indicators of scientific and technical inputs and outputs, these efforts have proved stillborn. Neither UNESCO nor the Economic Commission for Europe of the United Nations were able to overcome the formidable political and scientific problems when they attempted to produce common definitions and standards in the 1960s.

This has meant that in practice the attempt to make valid comparisons between East and West Europe or with the United States has been left to serious academic investigators and the intelligence services of the various powers. The intelligence services do not often publish their estimates and when they do, they must be regarded as highly suspect, as well as of low quality from a scientific point of view. In the military field there are obvious pressures to overstate the scale of "enemy" activity in order to justify a greater arms build-up. Much of the spurious "missile gap" propaganda which preceded Kennedy's election and that of Reagan was based on pseudo-statistics of this type.

This means that a very great responsibility devolves upon a handful of serious scholars in the West and East who are ready to accept the task of attempting to make independent and reasonably objective estimates and comparisons. Even without all the political and ideological pressures and confusion, this task would be difficult enough. There are genuine difficulties in the measurement of inputs into the science-technology system and in the definition of what constitutes "research and development" and what should be defined as "related scientific and technical activities". The West European countries have chosen to define R and D rather narrowly and the "Frascati" definitions of what constitutes R and D exclude such activities as geological exploration, information services, detail design and engineering or market research.

Largely following the Soviet definitions, the East European countries, including the GDR, measure a much wider range of scientific and technical activities and services. This has many great advantages and it is to be hoped that ultimately, when the present wave of false economies in government statistical services has blown over, the OECD countries will also measure these activities. In the meantime anyone who wishes to make serious comparisons between West and East Europe, as opposed to propaganda tricks, must find some way to adjust the East European statistics to the Frascati definitions by deducting the "non R and D" components. The Birmingham Centre pioneered in these estimates and with the unofficial cooperation of various East European scholars, the techniques have been greatly improved.

Building upon this experience Ray Bentley estimates that the GDR's industrial R and D effort has been lower, both absolutely and relatively, than that of the FRG. But his work goes far beyond this level of statistical comparison of inputs. Most of his book is devoted to the far more complex problem of "output" measurement, or to put it in a more straightforward way: the effectiveness with which scientific and technical resources are used to bring about technological change in the economic system.

Most studies of the Soviet Union and other East European countries as well as their own internal assessments, seem to agree that this is an area of serious relative weakness in the socialist countries. Whilst they have generally been more ready than comparable capitalist countries to commit resources to the finance of R and D and other scientific and technical activities, they have not yet found satisfactory ways for enterprises to take full advantage of these resources.

Much of the discussion of this weakness relates to the linkages between the specialised R and D units and the productive enterprises themselves. Another weakness is often supposed to be the linkage between those parts of the system whose main concern is with fundamental research and those responsible for applications and development. These problems are not confined to Eastern Europe; they are the subject of serious debate and attempts at reform and organisational improvement everywhere in the world. The many changes which Ray Bentley describes in his study of the GDR system would find some parallel in most other countries, and even in the internal R and D apparatus of many large multi-national firms.

Nevertheless it does seem to be true that the problems in Eastern Europe over the past thirty years or so have been particularly acute in this field. To understand why this might be is extremely important both for the citizens of these countries and for the rest of the world. That is why Ray Bentley's study is so much to be welcomed. It is an excellent scholarly survey of an extremely difficult problem and it deserves to be read by everyone with an interest in the problems of improving the performance of science-technology systems anywhere in the world.

Christopher Freeman
Professor of Science Policy
Science Policy Research Unit
University of Sussex

Acknowledgments

Although it was not possible to conduct this project direct in the GDR, I did have the fortunate possibility to be based in the Federal Republic of Germany. I would like to thank the Abteilung für Wissenschaftsforschung, and its director, Professor Helmut Baitsch, of the University of Ulm for the provision of guest research facilities. I would also like to thank the many individuals and organisations that endeavoured to help with information for the study, and in particular Thomas Werner, Hannelore Kutscher, Helga Kopp and colleagues of the Universitätsbibliothek Ulm; the librarians at the Universities of Marburg and Freiburg; and Viv Johnson and Val McNaughton of the SPRU library at the University of Sussex. For financial support, I am grateful for scholarships from the British Science Research Council and the German Academic Exchange Service (DAAD).

An earlier version of the study was accepted as a doctoral thesis by the University of Sussex. A significant impetus to the research derived from the extremely useful comments of my supervisor, Keith Pavitt, Senior Fellow at the Science Policy Research Unit, on the drafts of the thesis. His wide knowledge of Western research, development and innovation was particularly valuable in helping me to establish points of comparison and contrast between East and West. I also benefited greatly from comments on the thesis made by Dr. Ronald Amann of Birmingham University and Professor Christopher Freeman of Sussex University. Further helpful remarks were kindly contributed by Dr. Karen Worgen and Claire Booker on the manuscript of the book, and by Dr. Otto Keck and Professor Ina Spiegel-Rösing on some early draft chapters of the thesis. Moreover, I am grateful to Dean Birkenkamp of Westview Press for his efficient and friendly collaboration during the preparation of the book, and to Janet Barber, Jan Mathiasen, and Heidi Nieß for their ready and careful assistance with proofreading.

Most of all, I wish to thank Angela Erkert. Without her interest, her help, and her encouragement, this book would not have been possible. She also did an excellent and patient job typing the manuscript, and its earlier drafts, in a language which for the most part is not her mother tongue.

Finally, I must emphasise that the persons, institutions, and organisations mentioned above do not necessarily concur with the analysis and conclusions contained in the study. The contents and all errors of the book are entirely my responsibility.

Raymond Bentley

1 Introduction

THE ECONOMIC AND POLITICAL BACKGROUND

The German Democratic Republic prided itself in the mid-seventies on being one of the ten leading industrial nations of the world.[1] Its economic performance has sometimes been referred to as the "red economic miracle" and according to World Bank statistics for 1977, its per capita gross national product was the highest in Eastern Europe, higher than that of Great Britain and Italy, but lower than that of the Federal Republic of Germany.[2] Owing to the different methods of statistical accounting in East and West, and the problem of stipulating exchange rates, such comparisons must be treated with some caution. World Bank figures indicate that the GDR's per capita GNP in 1977 was 41 % lower than that of the FRG, whereas figures calculated by the German Institute for Economic Research in West Berlin show it to have been about 20 % lower for the period 1967-1976. This institute also estimated GDR industrial labour productivity to be 72 % that of the FRG for 1967, and 65 % for 1976.[3]

All Western comparisons of productivity in the two Germanies indicate a significant lag for the GDR, even though they differ with respect to the size of the lag. The existence of this productivity differential has been officially acknowledged by the GDR itself[4] and it is probably more a result of factors endogenous than exogenous to the economic system.[5]

One exogenous difference between the GDR and FRG is the energy and resource base. Most observers agree that this is inferior in the GDR. Only brown coal and potash are available in large quantities, and there is little in the way of hard coal, iron ore and other minerals. Another exogenous difference concerns the effects of war and reparations. East German industry suffered more war damage than that of West Germany, the country received no help equivalent to Marshall Aid, and it had to bear a greater reparations burden than the Western Zone. Factories were dismantled and transported to the Soviet Union, and up to 1954 the USSR drew on current production as part of the reparations payment. In West Germany fewer factories were dismantled and no reparations drawn on current production. Even after the period of reparations payment, it is possible that the GDR was obliged to export certain products eastwards which it needed domestically. The physicist Manfred von Ardenne complained in 1963, for instance, that:

> In many of our factories there is a danger of our production facilities becoming out of date. Often we export our most modern machines and forget the equipment of our own factories, where the same machines would bring a far higher gain by raising labour productivity and production quality.[6]

A third exogenous difference involves the greater disruption of production and trade for the GDR owing to the division of Germany. This severed the GDR's manufacturing industry from its traditional sources of raw materials, semi-manufactures and other products. Particularly serious was the fact that the metallurgical and heavy machine industries of Germany had been concentrated mainly in the Ruhr.

Yet even a poor endowment of natural resources is not necessarily an obstacle to economic success, and in the long term the effects of war damage, post-war disruptions, and obligations to the Soviet Union should, to an appreciable extent, have been overcome. Bearing in mind the fact that both Germanies are similarly endowed with human skills, have common traditions and culture, and were at a comparable level of industrialisation in 1945, it is unlikely that these exogenous factors were the main cause of the gap in productivity between the two Germanies evident in the sixties and seventies.

One factor which probably played a more important role and which had both exogenous and endogenous aspects was the heavy migration from the Eastern to the Western Zone, particularly of qualified persons. In part, this loss of manpower resulted from the strong economic performance of the FRG and the effect of Western propaganda, but in part it was a vote of protest against Stalinist policies and the economic system. By the time the Wall was built in 1961, the GDR had lost to the FRG more than two and a half million people, including approximately one-third of its persons possessing academic qualifications.

Other factors which played an important role were wholly or mainly endogenous. One was the style of economic management and organisation in the GDR. Another was the lower GDR participation in international trade relative to the FRG, and the strong orientation of the GDR's foreign trade towards Eastern Europe. In 1950, 72 % of the GDR's trade was conducted with the socialist/Comecon countries, 40 % with the USSR, 27 % with the capitalist countries, and 0.4 % with the developing countries. Twenty-five years later, the pattern was very similar.[7] The main disadvantages of this for the GDR compared with the FRG were a more restricted access to modern technology and a lower pressure of international competition. A third endogenous factor, closely related with the previous two and of central interest to this study, was that the GDR had a weaker capacity for technological change. GDR sources sometimes acknowledged that the country had a lower overall level of technology compared with the FRG yet, as will be argued in the remainder of this section, it is unlikely that the two Germanies attached very different degrees of importance to the goal of technological change.

Economists in both societies have long regarded technological change as a vital source of economic growth. Whereas certain sections of the West German population have sometimes expressed disquiet about the directions and consequences of science and technology, voices in the GDR sceptical or critical of the benefits held to ensue have been rare.

Indeed, technological change has traditionally occupied a positive central place in orthodox Marxist and Eastern European thought and ideology. Marx stated that competition forced the capitalist to improve the means of production and that large-scale industry depended on "the general level of science and the progress of technology, or the application of this science to production".[8] Eventually, the relations of production would hinder the development of the productive forces and result in crises and class conflict. Conditions would then become ripe for the overthrow of capitalism and the introduction of socialism. Lenin assigned science and technology a crucial role in the industrialisation and development of the Soviet Union. He conceded that transforming the rich inheritance of culture, knowledge and technology from a "tool of capitalism" into a "tool of socialism" would be no easy task, especially as it necessitated the assistance of "bourgeois specialists" and their conversion from "servants of capitalism" into "servants of the working masses".[9] Yet he remained optimistic: "not even the most sinister force will be able to withstand the alliance of science, proletariat and technology".[10]

Stalin, perceiving the spectre of war, stressed that a rapid rate of technological change was essential for the country's survival. The Soviet Union, he said, lagged behind the advanced countries by fifty to a hundred years. If this gap was not closed within the next ten years, the country would be "crushed". Although much had already been achieved in the construction of socialism, it was necessary for the bolsheviks to "learn technology", and "master science"; such matters could no longer be left mainly in the hands of bourgeois specialists. "Some say it is difficult to master technology. Wrong! There are no fortresses that the bolsheviks could not take."[11] In the event, the Soviet Union was not "crushed" though it suffered great human and economic loss; and despite the considerable progress made in "learning technology" and "mastering science" during Stalin's rule, the USSR's technological gap with the West was not closed in most sectors. Consequently Khrushchev, on consolidating power, took up his predecessor's aim of "catching up and overtaking" the USA in per capita production, and made it a central feature of his doctrine concerning the "peaceful competition of the two world systems".

Science was described as a "direct productive force" in the 1961 party programme of the Soviet Union, and afterwards Eastern European theoreticians exhibited increased interest in the relationship between science, technology and societal development. In the GDR this relationship found theoretical expression in concepts such as "scientific-technical progress", the "technical revolution" and the "scientific-technical revolution".[12] Judging from the titles of articles published in the party journal _Einheit_ between 1961 and 1975, the first concept seems to have been prevalent from 1961 to 1964 and from 1971 to 1975, the second from 1964 to 1966, and the third from 1967 to 1971. It is unlikely to be a coincidence that these changes in the official use of terminology took place roughly at the same time as changes in economic policy.

For the GDR's political leadership, economic success through technological change was a prerequisite for their legitimation: they hoped to win the allegiance of the population; to convince the Soviet Union that the GDR was both a politically reliable and technologically useful junior partner in Comecon; and not least, to demonstrate the "superiority of socialism over capitalism". The "competition of the two world systems" was more than just a theoretical postulate for the GDR, it was real. Unlike any of the other Eastern European states, the GDR shared its language, history and cultural tradition with a Western capitalist country. West German radio, television, relatives, and friends enabled the GDR citizen to compare conditions in the two Germanies. Clearly the task for the GDR was to "catch up and overtake" the FRG.[13]

> The outcome of the peaceful competition between the two world systems will mainly be determined by which societal order best promotes science and technology, and makes best use of the opportunities created.[14]

DATA, METHODS AND LIMITATIONS

The GDR has not made public any critical inquiry into its science and technology policy, and no substantive assessment of the GDR's capacity to innovate and diffuse technology has been published in the West.[15] Western scholars interested in the phenomenon of technological change under socialist conditions have, on the whole, chosen to research the situation in the Soviet Union.[16] For many observers, the GDR is one of the most difficult countries in Eastern Europe with regard to obtaining useful information: official agencies are usually uncooperative; publications are often verbose and dense with jargon; and important data and statistics tend to be fragmentary.

"Technological change" will be understood to comprise all those activities leading to and culminating in the introduction by an enterprise of a new or improved product or process. This concept is wider than the more usual Western definitions of "innovation" meaning the first ever commercial use of a new product or process, together perhaps with the events leading to this. Essential to raising technological levels and productivity is the diffusion of existing technology. The interest of this book is therefore not restricted to the ability of the GDR economy to bring forth new technology for the first time ever: attention is also paid to its capacity to stimulate diffusion.

An attempt is made in the study to uncover how the system of technological change in the GDR worked in actual practice and not merely how it was supposed to work. GDR evidence is compared with aspects of technological change in other countries, including the USSR, and where possible statistical comparisons are made,particularly with the FRG. The approach is also strongly historical to accentuate the dynamics, the interrelationships and the problems of the system of technological change. The year 1975 was imposed as an upper limit on the period investigated. One reason for this was that obtaining GDR documents and materials on a particular topic often takes a considerable amount of time, and it is virtually impossible to be completely up to date on all the topics dealt with in this study. Another was to benefit from historical perspective: each year the GDR publishes a vast number of announcements and decrees pertaining to its system of technological change, but only later is it possible to discern which of these were actually put into effect and how.

Four broad "policies" are distinguished in the book and it is necessary to stress that these are analytical constructs. In common with most other countries, the GDR has not had an explicit, self-contained and coherent policy for technological change. "Policies" will be understood to comprise not only measures aimed specifically at stimulating, but also measures affecting technological change. An investigation which concentrated solely on the former type of measures would fail to illuminate some very important influences on the rate of introducing new and improved technology in the GDR, many of which have been elements of more general economic, industrial, educational, and cultural policies.

It is also necessary to stress that for reasons of space, time and lack of data, many aspects of technological change must be omitted. Some obvious and significant omissions are details of branch strategies, the role of the banks, the valuation of capital stock, the production fund charge, the patent system, the transfer of technology from the West, the degree to which indigenous research and development is related to such technology, the role of the GDR in assimilating Western technology for Comecon, and other aspects of scientific-technological cooperation between the GDR and the rest of Comecon. There is also little formal detail about central agencies concerned with science and technology policy and about the R&D planning process: such information may be found in the book by Scherzinger.[17]

THE ORGANISATION OF THE BOOK

First a brief analysis is made of the technological levels of GDR industry. Chapter Two concentrates upon the country's three most important industrial sectors and, utilising trade and production statistics, compares the sophistication of GDR technology with that of the FRG, and to a lesser extent with that of the countries of the OECD and Comecon.

Chapter Three then studies the industrial R&D effort of the GDR from the sixties to the seventies. It examines the distribution of R&D manpower by type of establishment, the facilities and time available for R&D, the employment and qualification structure of R&D manpower, and the size and direction of the industrial R&D effort. Some experimental comparisons are also made with industrial R&D in the FRG and other countries.

The next four chapters analyse policies affecting technological change over the period 1945-1975. Chapters Four, Five and Six deal with the industrial sector, and Chapter Seven with the academic sector and its link with industry.

Chapter Four examines the policy of "detailed central planning" operative up to 1962. After giving a brief historical outline of central planning in the GDR, it seeks to identify and analyse some of the major obstacles to technological change under the planning system, and then discusses some factors which counteracted these obstacles to a certain degree.

In 1963 the party leadership introduced a new policy which aimed at stimulating greater industrial efficiency in general, and a higher rate of technological change in particular. This was the "New Economic System", and it involved substantial organisational reform together with the replacement of detailed planning procedures by financial instruments. Chapter Five attempts to analyse how far the NES succeeded in overcoming the obstacles to technological change characteristic of the previous system, and what the consequences were for technological change of the disbandment of the NES in 1971.

An important reason for the termination of the NES was the influence, under conditions of taut national output planning, of what might be considered to be an independent policy towards technological change launched in 1968. The aim was to overtake some of the current highest levels of technology in the world, and implicitly in the FRG, through the intensive promotion of selected priority areas. Chapter Six takes a closer look at this policy, designated here as the "offensive strategy", and focuses on its two main aspects: priorities for technological change, and concentration in industry.

Over the period 1945-1975, policy measures were implemented to couple the academic sector, which comprised the Academy of Sciences and the universities, with industry. Chapter Seven attempts to identify the main measures involved and looks more closely at aspects of the research connection between the two sides.

Chapter Eight suggests reasons why the GDR's capacity for technological change is weaker than that of the FRG, but stronger than that of the USSR. Finally some reflections are made towards a model of technological change in the GDR.

NOTES

1. GDR. 100 Questions 100 Answers (Berlin: Panorama Verlag DDR, 1975), 85; "GDR Engineering Week in Great Britain" (Advertisement), Financial Times (10 November 1975), 8.

2. World Bank Atlas 1979, quoted in: Der Fischer Weltalmanach 1981 (Frankfurt am Main: Fischer Verlag, 1980), 767.

3. DIW, "Das Sozialprodukt der DDR und der Bundesrepublik Deutschland im Vergleich", Wochenbericht, Nr. 23-24 (1977), 200.

4. 25 % was quoted by Walter Ulbricht in 1963 at the Sixth Party Congress of the SED. "Um die schnelle Steigerung der Arbeitsproduktivität auf der Grundlage höchster Leistungen von Wissenschaft und Technik", Neues Deutschland (16. Januar 1963), 4.

5. The following classification is influenced by P. Gregory and G. Leptin, "Similar Societies Under Differing Economic Systems: The Case of the Two Germanies", Soviet Studies, No. 4 (1977), 519-542.

6. M.A. Ardenne, Wege zur Steigerung der Weltmarktfähigkeit unserer industriellen Erzeugnisse (Berlin: Verlag Die Wirtschaft, 1963), 24.

7. In 1975 the percentages were: 70 % with the socialist countries; 66 % with Comecon; 36 % with the USSR; 26 % with the capitalist countries; and 4 % with the developing countries. Calculated from: St.J.DDR 1976, 265.

8. K. Marx, Grundrisse der Kritik der politischen Ökonomie (Berlin: Dietz Verlag, 1974), 592.

9. W.I. Lenin, "Rede auf dem I. Gesamtrussischen Kongress der Volkswirtschaftsräte, 26. Mai 1918", in Werke, Band 27 (Berlin: Dietz Verlag, 1972), 407-409.

10. W.I. Lenin, "Rede auf dem II. Gesamtrussischen Verbandstag des Medizinischen und Sanitätspersonals, 1. März 1920", in Werke, Band 30 (Berlin: Dietz Verlag, 1972), 394.

11. J. Stalin, "Über die Aufgabe der Wirtschaftler", in Fragen des Leninismus (Berlin: Dietz Verlag, 1951), 398-401.

12. For an account of the third concept in the Soviet Union see: J. Cooper, "The Scientific and Technical Revolution in the USSR", Co-existence, No. 2 (1981), 175-192.

13. See e.g. Akademie der Wissenschaften der DDR, Zentralinstitut für Geschichte, DDR - Werden und Wachsen (Berlin: Dietz

Verlag, 1975), 326; "Gesetz über den Siebenjahrplan zur Entwicklung der Volkswirtschaft der Deutschen Demokratischen Republik in den Jahren 1959 bis 1965", in W. Ulbricht, Der Siebenjahrplan des Friedens, des Wohlstandes und des Glücks (Berlin: Dietz Verlag, 1959), 159.

14. W. Ulbricht, Die Durchführung der ökonomischen Politik im Planjahr 1964 unter besonderer Berücksichtigung der chemischen Industrie (Berlin: Dietz Verlag, 1964), 53.

15. For some Western work see: V. Slamecka, Science in East Germany (New York: Columbia University Press, 1963); Der Bundesminister für wissenschaftliche Forschung, Bundesbericht Forschung II (Bonn, 1967), 128-135; W. Bergsdorf, "Produktionsfaktor Wissenschaft. Zur Wissenschafts- und Forschungspolitik in der DDR", Die neue Gesellschaft, Nr. 2 (1968), 167-175; T. Ammer, "Reform der Deutschen Akademie der Wissenschaften zu Berlin", Deutschland Archiv, Nr. 5 (1970), 546-551; P.R. Lücke, "Wissenschaft und Forschung im anderen Teil Deutschlands. Wissenschaftliche Akademien", Deutsche Studien, Nr. 32 (1970), 398-405; M. Rexin, "Die Entwicklung der Wissenschaftspolitik in der DDR", in Wissenschaft und Gesellschaft in der DDR (München: Carl Hanser Verlag, 1971), 78-121; A. Leisewitz and R. Rilling, "Wissenschafts- und Forschungspolitik in BRD und DDR", in BRD - DDR Vergleich der Gesellschaftssysteme (Köln: Pahl-Rugenstein Verlag, 1971), 365-384; W. Gruhn, "Zur Industrieforschung in der DDR. Teil I: Grundsätzliche Probleme", Deutsche Studien, Nr. 44 (1973), 368-377; and "Teil II: Die Struktur der Industrieforschung", Deutsche Studien, Nr. 45 (1974), 64-83; F. Oldenburg, "Zur Organisation der Forschung in der DDR. Bemerkung zur neuen 'Verordnung über die Leitung, Planung und Finanzierung an der Akademie der Wissenschaften und an Hochschulen' der DDR", Berichte des Bundesinstituts für Ostwissenschaftliche und Internationale Studien, Nr. 1 (1973); R. Rytlewski, "Zu einigen Merkmalen der Beziehungen der DDR und der UdSSR auf dem Gebiet der Naturwissenschaft und Technik", Deutschland Archiv, Sonderheft (1973), 55-60; Institut für Gesellschaft und Wissenschaft Erlangen (ed.), Wissenschaft in der DDR (Köln: Verlag Wissenschaft und Politik, 1973), igw Informationen. Zur Wissenschaftsentwicklung und -politik in der DDR, and abg. Analysen und Berichte aus Gesellschaft und Wissenschaft; M. Usko, Hochschulen in der DDR (Berlin: Verlag Gebr. Holzapfel, 1974); A. Scherzinger, Zur Planung, Organisation und Lenkung von Forschung und Entwicklung in der DDR (Berlin: Duncker und Humblot, 1977). Some interesting ideas on technological change in the GDR are also contained in: D. Granick, Enterprise Guidance in Eastern Europe. A Comparison of Four Socialist Economies (Princeton, N.J.: Princeton University Press, 1975); D. Granick, "Variations in Management of the Industrial Enterprise in Socialist Eastern Europe", in Reorientation and Commercial Relations of the Economies of Eastern Europe, A Compendium of Papers submitted to the Joint Economic Committee Congress of the United States (Washington, D.C.: U.S. Government Printing Office, 1974), 229-247.

16. For some good examples of this work see: R. Amann, M.J. Berry and R.W. Davies, "Science and Industry in the USSR", in E. Zaleski et al., *Science Policy in the USSR* (Paris: OECD 1969), 378-557; J.S. Berliner, *The Innovation Decision in Soviet Industry* (Cambridge, Mass.: The MIT Press, 1976); R. Amann, J. Cooper and R.W. Davies (eds.), *The Technological Level of Soviet Industry* (New Haven and London: Yale University Press, 1977); R. Amann and J. Cooper (eds.), *Industrial Innovation in the Soviet Union* (New Haven and London: Yale University Press, 1982). For a study of "technological progress in Czechoslovakia and Austria, see: H.G.J. Kosta, H. Kramer and J. Slama, *Der technologische Fortschritt in Österreich und in der Tschechoslowakei* (Wien: Springer Verlag, 1971).

17. Scherzinger, *op. cit.*, note 15.

2
The Technological Background

THE QUANTITATIVE APPRAISAL OF GDR TECHNOLOGY

The declarations issued by GDR leaders and the assertions of Western observers and journalists all point to a lower level of technology in the GDR than the FRG, yet only rarely have these general statements been accompanied by specific details. Hardly anything has been published in the way of a quantitative assessment of GDR technology. A survey of West German and British manufacturers, trading organisations and ministries, undertaken for this study, found that surprisingly little appears to be known in the West.[1] While it would clearly exceed the scope of this chapter to undertake a thorough analysis of technological levels in the GDR relative to the FRG and other countries, a brief assessment may be made.[2] The method adopted is to examine aspects of foreign trade and production with regard to the three main industrial sectors of GDR industry. These sectors, in terms of production output, annual growth rate, exports and research intensity, were the machine and vehicle building industry; the electrotechnical, electronics, and instrument building industry; and the chemical industry.

One rough indicator of technological levels is the composition of trade between the two Germanies. The assumption is that a country strong in research, innovation and diffusion will have a fairly high proportion of sophisticated technology in its exports. Table 2.1 gives a breakdown by industrial sector of trade between the two Germanies for the years 1976 and 1978. The trade pattern in both years was very similar and suggests that the GDR has been exporting less sophisticated machine building, electrotechnical and chemical products to the FRG than it has been importing in return. The three main export categories

Table 2.1
Inter-German Trade

Most Important GDR Exports to FRG

		Percentage of Total Exports 1976	1978
1	Textiles, Clothing	19.5	18.7
2	Agricultural Products	18.3	14.3
3	Mineral Oil Products	14.8	14.6
4	Chemical Products	8.3	9.1
5	Machines, Electrotechnical Products	7.6	7.2
6	Iron and Steel, Products of Drawing Plants and Cold-Rolling Mills	4.9	3.9

Most Important GDR Imports from FRG

		Percentage of Total Imports 1976	1978
1	Machines, Electrotechnical Products	23.5	31.3
2	Chemical Products	19.3	15.7
3	Iron and Steel, Products of Drawing Plants and Cold-Rolling Mills	10.3	7.0
4	Agricultural Products	6.7	7.9
5	Textiles, Clothing	5.5	4.8
6	Crude Mineral Oil	5.4	5.1

Source:

Based on Presse- und Informationsamt der Bundesregierung, *Bulletin* Nr. 132 (10. Dezember 1976), 1263. Bundesminister für innerdeutsche Beziehungen, *Information* Nr. 5 (1979), 12.

Note:

(1) The figures for 1976 are for the first nine months.
(2) In 1978 the product group "wooden goods" accounted for 4.9 % of the GDR's exports to the FRG. It thus displaced "iron and steel" from position six.

of the GDR to the FRG were textiles and clothing, agricultural products, and mineral oil products. Those of the FRG to the GDR were machines and electrotechnical products, chemical products, and iron and steel.

In the more disaggregate examination of the GDR's three main industrial sectors which follows, the indicator used most is the kilogram-price of exports from the two Germanies on third markets. Kilogram-prices have been an instrument of foreign trade analysis for a number of years in Eastern Europe, especially Czechoslovakia, and some scholars in the West have adopted the method to study Soviet technology.[3] The basic assumption of the method is that products having a high kilogram-price embody more sophisticated technology.

It should be noted that kilogram-prices can also reflect the influence of other factors such as inflation, dumping, the quality of commercial work, consumer prejudice, and the structure of the product group represented in the trade statistics.[4] In this assessment it is assumed that GDR products are sold abroad at current prices and are also affected by inflation. Furthermore, kilogram-prices for GDR products are compared with those of the FRG and other countries for a particular year: the interest is in relative kilogram-prices within a year rather than the development of absolute kilogram-prices over a period of time. Dumping usually occurs when there is a glut on the world market with regard to traditional, labour-intensive goods and it is unlikely to have applied to the three sectors under review. Conceivably the GDR's political isolation from the Western world up to the seventies caused her commercial work to be inferior to that of the FRG, and her products to be subject to Western, and especially West German, consumer prejudice. This difficulty may be avoided to some extent by examining in the first instance the relative GDR performance on the Yugoslavian market: owing to the weakness of its currency and its industry, Yugoslavia was inclined to encourage trade with the countries of Comecon. Lastly, product groups can be extremely heterogeneous. Reference to the Standard International Trade Classification (SITC) shows, for example, that the group "machinery and transport equipment" (SITC 7) covers an assortment of products ranging from textile machines to radios, and these will have very different prices per weight. To minimise this problem, kilogram-prices are calculated for the subdivisions of the SITC groups where the data permit. When only very small quantities were involved, they are excluded from the investigation as these, in all probability, were special orders.

The principal year chosen for the kilogram-price comparisons is 1968. This is appropriate since the only detailed statistics which could be obtained on the GDR's industrial research and development relate to the years 1963/64, and it will be of interest in the next chapter to enquire whether products obtaining comparatively low kilogram-prices had previously been characterised by relatively low R&D efforts. Moreover,

1968 was the year in which the GDR launched an ambitious programme for technological change, and it comes towards the end of the period surveyed in this book: by this date the immediate postwar disruptions and the effect of reparations should largely have been overcome. For each of the industrial sectors under review, the Germanies are compared first on the Yugoslavian market. Then the most competitive GDR products of this market are compared with their FRG counterparts on the "harder" market of the Organisation for Economic Co-operation and Development (OECD). Supplementary kilogram-price comparisons of the GDR with other Comecon countries and with the OECD are also made.

Two other indicators of technological levels are used in the examination of the three industrial sectors and might be subsumed under the heading "aspects of production". One is the comparison of the output per capita of certain "modern" products: in the electronics industry, transistors are more modern than valves, and in the chemical industry, synthetic fibres are more modern than viscose fibres. Such newer generations of products are usually more research intensive than the older ones, and involve more complex manufacturing processes. The other indicator used, which applies particularly to the organic chemical industry, is the structure of the raw material base. After the Second World War, petroleum and natural gas rapidly became the main raw material for the organic chemicals, and especially the synthetic materials industry in the leading capitalist countries. This enabled dramatic economies in production costs to be made.[5]

THE MACHINE AND VEHICLE BUILDING INDUSTRY

Machine and vehicle building accounted for about one quarter of the GDR's industrial production in 1975,[6] and for roughly 42 % of its exports.[7] Table 2.2 compares the kilogram-price for GDR machine and vehicle building exports to Yugoslavia in 1968 with those for the FRG. The GDR's average kilogram-price amounted to 0.74 that of the FRG, and for only eight of the thirty-two items listed did the GDR achieve a kilogram-price greater than or equal to that of the FRG. This suggests that on the whole, the technological level of the machine and vehicle building industry in the GDR lagged behind that of the FRG. Comparatively low kilogram values (less than 0.51 those of the FRG) were obtained by the GDR items: "textile machinery"; "excavating, levelling, and boring machinery"; "heating and cooling equipment"; "machine tools for

Table 2.2
Ratio of Kilogram-Prices Obtained by GDR and FRG on Yugoslavian Market in 1968 — Machine and Vehicle Building

SITC	Description	GDR : FRG
711	Power Generating Machinery, other than Electric	0.66
711.5	Internal Combustion Engines, not for Aircraft	1.51
712	Agricultural Machinery and Implements	0.52
714	Office Machines	1.07
714.1	Typewriters and Cheque-Writing Machines	0.98
714.2	Calculating and Accounting Machines	0.73
715.1	Machine Tools for Working Metals	0.84
715.2	Other Metalworking Machinery	2.04
715.22	Rolling Mills and Rolls, for Metalworking	2.63
717.1	Textile Machinery	0.49
717.3	Sewing Machines	0.59
718.1	Paper Mill and Pulp Mill Machinery	0.94
718.2	Printing and Bookbinding Machinery	0.71
718.3	Food-Processing Machinery, excluding Domestic	1.23
718.4	Construction and Mining Machinery, n.e.s.	0.46
718.42	Excavating, Levelling, Boring etc. Machinery	0.46
718.5	Mineral Crushing etc. and Glass-Working Machinery	1.05
719.1	Heating and Cooling Equipment	0.50
719.15	Refrigerators, not Domestic	0.65
719.2	Pumps and Centrifuges	0.75
719.3	Mechanical Handling Equipment	0.76
719.5	Powered Tools, n.e.s.	0.83
719.52	Machine Tools for Working Wood, Plastics, etc.	0.45
719.6	Other Non-Electrical Equipment	0.76
719.7	Ball, Roller or Needle Roller Bearings	1.23
719.8	Machinery and Mechanical Appliances, n.e.s.	0.53
719.9	Parts and Accessories of Machinery, n.e.s.	0.74
731	Railway Vehicles	3.04
732	Road Motor Vehicles	0.79
732.1	Passenger Motor Cars, other than Buses	0.74
732.4	Special Purpose Lorries, Trucks, Vans	0.46
732.9	Motorcycles and Parts	0.66
Total		0.74

Source: Calculated from statistics in: *Trade by Commodities. Market Summaries. Imports January - December 1968, Series C* (Paris: OECD).

Note: n.e.s. = not elsewhere specified.

working wood, plastics etc."; and "special purpose lorries, trucks and vans". Table 2.3 compares the kilogram-prices of the GDR's eight "leading" products with those of the FRG on the "harder" OECD market. Here the GDR equalled the FRG for only one item: "mineral crushing and glass working machinery".

The GDR would fare better in comparisons of technological levels with certain other Western countries than with the FRG, for the latter is a rigorous yardstick. Table 2.4 compares kilogram-prices for the GDR and some Western countries on the OECD market, the SITC positions being those for which the GDR achieved kilogram values at least 0.80 as high as the FRG in Yugoslavia. It shows that the GDR obtained values equal to or higher than Italy in one case, Denmark and Britain in two cases, the Netherlands in three cases, Norway in four cases, Austria in five cases, and Belgium/Luxemburg in six cases. Statistics on world machine tool production and trade also show that the FRG in 1968 was the world's leading exporter and the second largest producer. If one accepts their

Table 2.3
Ratio of Kilogram-Prices for GDR and FRG
Machine Building Products on OECD Market in 1968

SITC	Description	GDR : FRG
711.5*	Internal Combustion Engines, not for Aircraft	0.52
714	Office Machines	0.54
715.2	Other Metalworking Machinery	0.85
715.22	Rolling Mills and Rolls, for Metalworking	0.62
718.3	Food-Processing Machinery, excluding Domestic	0.40
718.5	Mineral Crushing etc. and Glass-Working Machinery	1.00
719.7	Ball, Roller or Needle Roller Bearings	0.86
731	Railway Vehicles	0.78

Source: As Table 2.2.

*Figures for the GDR's main OECD customer (Italy) are used because data for the total OECD market are not available.

Table 2.4
Relative Kilogram-Prices for GDR and Certain Western Countries—Machine Building Products on OECD Market in 1968

SITC	Description	GDR/FRG	GDR/I	GDR/DK	GDR/GB	GDR/NL	GDR/N	GDR/A	GDR/B/L
711.5*	Internal Combustion Engines, not for Aircraft	0.52	-	0.40	1.04	0.61	-	0.67	0.29
714	Office Machines	0.54	0.54	1.08	0.55	0.58	1.14	1.62	0.95
714.1	Typewriters and Cheque-Writing Machines	0.56	0.47	-	0.41	0.34	-	-	-
715.1	Machine Tools for Working Metals	0.57	0.64	0.84	0.58	0.68	0.36	0.68	0.73
715.2	Other Metalworking Machinery	0.85	1.12	0.20	2.01	2.73	1.92	1.13	2.25
715.22	Rolling Mills and Rolls for Metalworking	0.62	0.74	-	0.90	1.73	-	1.30	1.28
718.1	Paper Mill and Pulp Mill Machinery	0.62	0.79	0.71	0.65	0.93	0.82	0.96	1.36
718.3	Food-Processing Machinery, excluding Domestic	0.40	0.40	0.38	0.53	0.46	0.72	0.49	0.49
718.5	Mineral Crushing etc. and Glass-Working Machinery	1.00	0.93	0.81	0.99	0.56	1.28	1.87	1.02
719.5	Powered Tools, n.e.s.	0.53	0.82	0.69	0.44	0.41	0.49	0.83	0.47
719.7	Ball, Roller or Needle-Roller Bearings	0.86	0.93	1.22	0.89	0.91	-	1.20	4.08
731	Railway Vehicles	0.78	0.96	0.44	0.60	1.70	1.21	0.39	1.09

Source: As Table 2.2.

Note: I = Italy, DK = Denmark, GB = Great Britain, NL = Netherlands, N = Norway, A = Austria, B/L = Belgium and Luxemburg.

*Figures for the GDR's main OECD customer are used because data for the total OECD market are not available.

validity for countries having controlled currencies, they show that the GDR was the third largest exporter after the FRG and USA, and the ninth largest producer, just behind France but ahead of Switzerland. Between 1971 and 1975 the GDR had export figures similar to those of the USA, and in 1972 actually occupied the number two position for world exports. Both countries, however, followed the FRG at a considerable distance, their export sales being about 27 % those of West Germany.[8]

Aggregate kilogram-price comparisons suggest that the machine and vehicle building industry of the GDR was the strongest in Eastern Europe. Table 2.5 shows that the GDR attained higher kilogram values than the other Comecon countries in each of the years 1961, 1965 and 1968. In 1961 it had also been ahead of the capitalist countries, though by 1965 had lost this lead possibly due to the greater liberalisation of the Yugoslavian market.

Table 2.5
Relative Kilogram-Prices for Machine and Vehicle Building on Yugoslavian Market (SITC 71 and 73)

	1961	1965	1968
GDR	1.32	0.99	0.84
Poland	0.53	0.44	0.51
Czechoslovakia	1.30	0.56	0.49
Hungary	0.78	0.77	0.37
USSR	0.67	0.50	0.61
EEC	0.96	1.00	0.95
World	0.98	0.81	0.85
OECD	1.00	1.00	1.00

Source:

Calculated from data in Table C, Appendix, of J. Slama and H. Vogel, "Niveau und Entwicklung der Kilogrammpreise im Außenhandel der Sowjetunion mit Maschinen und Ausrüstungen", Jahrbuch der Wirtschaft Osteuropas, Band 2 (1971), 443-483.

THE ELECTROTECHNICAL, ELECTRONICS AND INSTRUMENT BUILDING INDUSTRY

In 1975 this sector accounted for about 11 % of industrial production[9] and 8 % of exports.[10] Table 2.6 compares the kilogram-prices achieved by GDR and FRG electrotechnical, electronic and instrument building products on the Yugoslavian market in 1968. GDR products of this sector appear to have been considerably less competitive than those of the machine and vehicle building industry. The GDR's average kilogram-price was only 0.32 that of the FRG, and for none of the positions listed did the GDR have a kilogram-price equal to or greater than the FRG. The data suggest

Table 2.6
Ratio of Kilogram-Prices Obtained by GDR and FRG on Yugoslavian Market in 1968 —Electrotechnical, Electronics and Instrument Building

SITC	Description	GDR : FRG
722.1	Electric Power Machinery	0.36
722.2	Apparatus for Electrical Circuits	0.47
724.2	Radio Broadcast Receivers	0.54
725	Domestic Electrical Equipment	0.71
725.01	Domestic Refrigerators, Electrical	0.90
725.02	Domestic Washing Machines	0.88
725.05	Electric Space Heating Equipment	0.44
729	Other Electrical Machinery and Apparatus	0.10
729.4	Automotive Electrical Equipment	0.65
729.5	Electrical Measuring and Controlling Instruments	0.36
729.92	Electrical Furnaces, Welding and Cutting Apparatus	0.74
891.1	Phonographs, Tape and Other Sound Recorders	0.34
861	Scientific, Medical, Optical, Measuring and Control Instruments	0.50
861.6	Photographic and Cinematographic Equipment, n.e.s.	0.43
861.7	Medical Instruments, n.e.s.	0.61
861.9	Measuring, Controlling and Scientific Instruments	0.43
862.4	Photographic Film etc.	0.68
Total		0.32

Source: As Table 2.2.

two particularly important weaknesses of this sector in the GDR: scientific and other measuring instruments (SITC 861 and 729.5), and electronics (SITC 722.2, 724.2, 891.1). The latter point is also illustrated in Table 2.7, which shows that the GDR had a significantly lower production output of transistors per capita than West Germany. By implication, a third important weakness of this sector in the GDR was the computer industry.[11]

Within Comecon the GDR seems to have been the technological leader in instrument building. Table 2.8 shows that the GDR achieved higher kilogram-prices than the other main Comecon countries for the SITC items 861, 861.6 and 729.5. The GDR does not appear to have had a technological lead in the electrotechnical and electronic branches: only for the item "electric space heating" (SITC 725.05) did it have the highest kilogram-price. It was, however, one of the main producers in Comecon of electrotechnical and electronic goods. In 1968 the GDR had the highest per capita output of the Comecon countries for radios, domestic refrigerators and vacuum cleaners, and came second after the USSR for domestic electrical washing machines. (It was also ahead of the UK and Austria in the per capita output of radios, refrigerators and washing machines, and exceeded France with regard to vacuum cleaners.[12]) The GDR came fourth in Comecon after Hungary, Czechoslovakia, and the USSR for the production per capita of television sets although earlier in the sixties it had occupied the first position.[13]

Table 2.7
Production of Transistors in GDR and FRG, 1960-1975
(Number per Capita)

	GDR	FRG	GDR : FRG
1960	0	0.415	0
1965	0.298	1.617	0.18
1970	1.540	7.115	0.22
1973	3.422	7.862	0.43
1974	3.556	9.391	0.38
1975	3.606	5.258	0.69

Source:

Calculated from St.J.DDR 1976, 121 and 31; Statistische Praxis, Nr. 7 (1974), 4; St.J.BRD 1978, 50; and data supplied by Statistisches Bundesamt Wiesbaden from: Produzierendes Gewerbe. Fachserie 4 Reihe 3.1. Produktion im produzierenden Gewerbe nach Waren und Warengruppen.

Table 2.8
Relative Kilogram-Prices for Electrotechnical, Electronic and Instrument Building Products on Yugoslavian Market in 1968

SITC	Description	GDR	Poland	Czecho-slovakia	Hungary	USSR	OECD
722.1	Electric Power Machinery	0.56	0.73	0.58	0.53	0.92	1.00
722.2	Apparatus for Electrical Circuits	0.56	0.25	0.62	0.31	0.40	1.00
724.2	Radio Broadcast Receivers	0.51	-	0.50	0.18	0.88	1.00
725	Domestic Electrical Equipment	1.33	1.41	1.28	1.39	1.05	1.00
725.01	Domestic Refrigerators, Electrical	0.92	1.13	1.09	1.05	0.78	1.00
725.02	Domestic Washing Machines	1.51	-	1.69	1.70	-	1.00
725.05	Electric Space Heating Equipment	1.22	0.48	-	0.89	-	1.00
729	Other Electrical Machinery and Apparatus	0.19	0.22	0.80	1.12	0.70	1.00
729.2	Electric Lamps	0.35	0.19	0.36	0.42	0.27	1.00
729.3	Thermionic Valves and Tubes, Transistors	0.83	-	0.64	1.02	0.42	1.00
729.4	Automotive Electrical Equipment	0.64	0.45	0.78	-	0.52	1.00
729.5	Electrical Measuring and Controlling Instruments	0.47	0.18	0.22	-	0.39	1.00
729.92	Electrical Furnaces, Welding and Cutting Apparatus	0.73	0.85	0.54	0.40	0.62	1.00
891.1	Phonographs, Tapes and Other Sound Recorders	0.37	-	0.32	-	1.71	1.00
861	Scientific, Medical, Optical, Measuring and Control Instruments	0.53	0.36	0.39	0.43	0.41	1.00
861.6	Photographic and Cinematographic Equipment, n.e.s.	0.38	0.19	0.35	-	0.21	1.00
861.9	Measuring, Controlling and Scientific Instruments	0.44	0.35	0.39	0.48	0.35	1.00

Source: As Table 2.2.

THE CHEMICAL INDUSTRY

This industry accounted for about 15 % of the GDR's industrial production in 1975,[14] and together with building materials for 12.5 % of exports.[15] The GDR's average kilogram-price for chemicals on the Yugoslavian market was 0.94 that of the FRG, and at first sight this might suggest a rather similar level of technology for the two Germanies. Inspection of the kilogram-prices for the individual SITC subdivisions, however, reveals otherwise. Table 2.9 suggests that the GDR was strong in more "traditional" chemical products: nitrogen function compounds, inorganic chemicals, and soaps, cleansing and polishing preparations, but relatively weak in the more "progressive" branches of chemical technology: synthetic fibres (SITC 266), medicaments (SITC 541.7), and plastics (SITC 581.1). Table 2.10 compares the kilogram-prices of the "leading" GDR products with their FRG counterparts on the OECD market. Only for one item, "surface acting agents and washing preparations", did the GDR's kilogram-price approach that of the FRG.

Table 2.9
Ratio of Kilogram-Prices Obtained by GDR and FRG on Yugoslavian Market in 1968—Chemicals

SITC	Description	GDR : FRG
231.2	Synthetic Rubber and Rubber Substitutes	0.39
266	Synthetic and Regenerated Artificial Fibres	0.41
512	Organic Chemicals	0.71
512.1	Hydrocarbons and their Derivatives	0.70
512.2	Alcohols, Phenols, Phenol-Alcohols, Glycerine	0.78
512.5	Acids and their Halogenated Derivatives	0.67
512.7	Nitrogen Function Compounds	1.02
512.8	Organo-Inorganic and Heterocyclic Compounds	0.78
513	Inorganic Chemicals	1.24
514	Other Inorganic Chemicals	0.55
541.7	Medicaments	0.17
554	Soaps, Cleansing and Polishing Preparations	1.40
554.2	Surface-Acting Agents and Washing Preparations	1.39
561.1	Nitrogenous Fertilisers and Materials, n.e.s.	0.93
581.1	Products of Condensation, Polycond. and Polyaddition	0.43
581.2	Products of Polymerisation and Copolymerisation	0.87
Total		0.94

Source: As Table 2.2.

Table 2.10
Ratio of Kilogram-Prices for GDR and FRG—
Chemical Products on OECD Market in 1968

SITC	Market	Description	GDR : FRG
512.7	Italy	Nitrogen-Function Compounds	0.36
513	Italy	Inorganic Chemicals	0.72
554	OECD	Soaps, Cleansing and Polishing Preparations	0.87
554.2	OECD	Surface-Acting Agents and Washing Preparations	0.94

Source: As Table 2.2.

The inference that the GDR lagged behind in the more progressive branches of chemicals is also supported by output statistics. Table 2.11 compares the output of selected chemical products in the two Germanies between 1960 and 1975. The GDR had a relatively high output of synthetic rubber, viscose fibres, ammonia, carbide, soap and fertilisers, and a noticeably low comparative output of synthetic fibres and synthetic materials/plastics.

The relatively high figure for carbide indicates a GDR lag in changing over to petroleum and natural gas as the main raw material base for the organic chemical industry. This had been one of the major aims of the "Chemical Programme" launched in 1958, though by 1975 only about 60 % of the GDR's organic chemical output was derived from petroleum and natural gas as compared with around 95 % for the FRG. The figure for the GDR in 1975 corresponded to that of the FRG in 1963.[16] The drastic rise in oil prices may have played a role in later years for the GDR's slow changeover to oil as a feedstock. Other reasons were probably the high investment costs involved and, as in the USSR, technological conservatism.[17]

The technological level of the GDR's chemical industry would compare more favourably with those of other countries than the FRG, as West Germany is a world leader in chemicals. In 1970 the per capita kilogram output of plastics and synthetic resins was 70.7 in the FRG compared with 49.3 in Japan, 39.8 in the USA, and 31.1 in Italy. The GDR, with the value 20.2, was close behind Great Britain (24.2) and France (22.6), and ahead of Australia (17.9) and Canada (13.3).

Table 2.11
Ratio of Per Capita Production of Selected Chemical Products in GDR and FRG, 1960-1975

	GDR : FRG			
	1960	1970	1974	1975
Synthetic Rubber	3.39	1.32	1.37	1.70
Synthetic Fibres	0.48	0.36	0.52	0.67
Viscose Fibres	2.29	2.79	4.96	7.96
Methanol	0.48	0.68	0.67	1.15
Sulphuric Acid	0.74	0.89	0.70	0.89
Hydrochloric Acid	0.97	0.43	0.41	0.52
Ammonia	1.00	0.93	1.18	1.70
Sodium Hydroxide	1.35	0.89	0.55	0.67
Carbide	2.71	5.07	7.70	9.81
Organic Dyes	0.48	0.46	0.41	0.70
Soap	1.16	1.04	0.85	0.85
Heavy-Duty Detergents	-	-	0.59	0.55
Nitrogenous Fertilisers	0.97	0.89	1.04	1.37
Phosphatic Fertilisers	0.64	1.68	1.59	2.15
Potash Fertilisers	2.71	3.75	4.04	5.04
Synthetic Materials/Plastics	0.39	0.32	0.29	0.44
Tyres for Vehicles	0.42	0.29	0.37	0.48

Source: Calculated from data in "Zur Lage der Chemiewirtschaft in der DDR", *Wochenbericht*, Nr. 49 (1976), 458.

The per capita kilogram output of synthetic fibres in the same year was 8.1 in the FRG, 9.1 in Japan, 7.9 in the USA, 6.3 in Great Britain, and 4.5 in Italy. The GDR had the value 3.1, and was just behind Canada (3.4) and France (3.2).[18]

Within Comecon, the GDR's chemical industry seems to have occupied the leading position. It had the highest figure for per capita kilogram output of plastics and resins in 1970 and was followed by Czechoslovakia (16.8), Bulgaria (10.5), Romania (10.2), Poland (8.2), the USSR (6.9), and Hungary (5.3). It also had the highest figure for per capita kilogram output of synthetic fibres, being followed by Bulgaria and Czechoslovakia (2.1), Poland (1.7), Romania (1.4), the USSR (0.7), and Hungary (0.5).[19]

SUMMARY

This brief appraisal of three industrial sectors suggests that GDR technology has lagged behind that of the FRG, but has been of a leading level in Eastern Europe, and of a respectable level with regard to machine building and chemicals in the world as a whole. Relative to West Germany, the GDR seems to have been particularly weak in the more progressive branches of technology, such as instrument building, electronics, data processing, synthetic fibres, and plastics. Relative to Comecon, the GDR seems only in electrotechnology and electronics not to have had a technological lead. Whilst these propositions are consistent with scattered comments in GDR publications, and qualitative information obtained in the survey of Western firms and agencies, it is emphasised that firmer and more differentiated conclusions would necessitate a much more detailed study, using more data and more measures.

NOTES

1. For some brief Western material on GDR industry see the various entries in DDR Handbuch (Köln: Verlag Wissenschaft und Politik, 1979). See further: "Zur Lage der Chemiewirtschaft in der DDR", Wochenbericht, Nr. 49 (1976), 457-460, and K. Krakat, "Der Weg zur dritten Generation. Die Entwicklung der EDV in der DDR bis zum Beginn der siebziger Jahre", FS Analysen, Nr. 7 (1976).

2. For an outstanding and detailed assessment of technological levels in Soviet industry see: R. Amann, J. Cooper and R.W. Davies (eds.), The Technological Level of Soviet Industry (New Haven and London: Yale University Press, 1977).

3. See e.g. J. Slama and H. Vogel, "Niveau und Entwicklung der Kilogrammpreise im Aussenhandel der Sowjetunion mit Maschinen und Ausrüstungen", Jahrbuch der Wirtschaft Osteuropas, Nr. 2 (1971), 443-483; R. Amann and J. Slama, "The Organic Chemicals Industry of the USSR: A Case-Study in the Measurement of Comparative Technological Sophistication by Means of Kilogram-Prices", Research Policy, 5 (1976), 302-326.

4. The following remarks on these factors are heavily influenced by the discussions in Amann and Slama, op. cit., note 3, and Slama and Vogel, op. cit., note 3.

5. See the chapter on the Soviet chemical industry in Amann, Cooper, and Davies, op. cit., note 2.

6. Deutsches Institut für Wirtschaftsforschung, Handbuch DDR-Wirtschaft (Reinbek bei Hamburg: Rowohlt Verlag, 1977), Table 18 on p. 313.

7. Exact figures are not available. Estimate based on St.J.DDR 1976, 264 and 281.

8. The statistics are contained in various January and February issues of _American Machinist_ from 1970 to 1977. The main problem underlying these statistics is that of using realistic exchange rates. In the figures for 1968 the controlled currencies were generally converted at rates suggested by the governments of the countries involved (sometimes different from the "official rate"). In the figures for 1971 to 1975, conversion rates for the GDR were the official rates: effective commercial rates can be different. Changes in the relative values of free-market currencies also complicate the making of comparisons.

9. Deutsches Institut für Wirtschaftsforschung, _op. cit._, note 6, 313.

10. Estimate based on _St.J.DDR 1976_, 264 and 281.

11. This inference is supported by Krakat, see note 1.

12. _St.J.DDR 1970_, 49* and 50*.

13. _St.J.DDR 1964_, 34*; _St.J.DDR 1967_, 49*.

14. Deutsches Institut für Wirtschaftsforschung, _op. cit._, note 6, 313.

15. _St.J.DDR 1970_, 264.

16. "Zur Lage der Chemiewirtschaft in der DDR", _Wochenbericht_, Nr. 49 (1976), 459.

17. Amann, Cooper, and Davies, _op. cit._, note 2, 250.

18. J. Wilzynski, _Technology in Comecon_ (London: Macmillan, 1974), 192 and 194. Figures for 1974 showing a similar distribution may be found in _St.J.DDR 1976_, 52*.

19. _Ibid._

3
The Industrial Research and Development Effort

Comparatively little is known about the organisation, strengths and weaknesses of industrial research and development in the GDR. Data and materials on this subject are particularly fragmentary, and no series of R&D statistics has been published in the GDR approaching the scope of those issued by the West German "Stifterverband für die Deutsche Wissenschaft" or the Federal Government in its regular "Research Report". The aim of this chapter is to present and evaluate such evidence as could be collected on the following aspects of the GDR's industrial R&D effort: the distribution of manpower by type of establishment; facilities; the time spent on R&D; the employment and qualification structure of manpower; and the scale and direction of the R&D effort. Whilst the Academy of Sciences and the institutions of higher education also contributed to industrial R&D, these are discussed later in Chapter Seven. It is necessary to stress that in places recourse to assumptions and estimates cannot be avoided. To facilitate the evaluation some comparisons with the FRG and other countries are made, but because different statistical concepts are involved, adjustments to the data are necessary to render them comparable.

DISTRIBUTION OF INDUSTRIAL R&D MANPOWER BY TYPE OF ESTABLISHMENT

Research and development was performed in two broad kinds of institution in GDR industry: enterprise establishments and institutes. The term "institute" is used here to refer to the "central institutes", "branch institutes", "central development and design bureaux", "central laboratories", and "enterprise R&D establishments functioning as scientific-technical centres". A breakdown of these for 1963 is given in Table 3.1. The five "central institutes" were under

Table 3.1
Institutes in 1963

5	Central Institutes
	Foundry Practice, Leipzig
	Production Engineering, Karl-Marx-Stadt
	Welding Practice, Halle
	Automation, Dresden
	Light Metal Engineering, Dresden
50	Industrial Branch Institutes
21	Central Development and Design Bureaux
3	Central Laboratories
	Plastics Processing, Leipzig
	Electrical Appliances, Karl-Marx-Stadt
	Radio and Television, Dresden
12	Enterprise R&D establishments functioning as Scientific-Technical Centres (STCs)

Source:

K. Heuer, "Die wirtschaftliche Rechnungsführung in den naturwissenschaftlichen Instituten der sozialistischen Industrie und die Gewährleistung einer neuen Qualität der Planung, Leitung, Organisation und Wirtschaftsführung im wissenschaftlich-technischen Zentrum der VVB", *Wissenschaftliche Zeitschrift der Humboldt-Universität zu Berlin*, Mathematisch-Naturwissenschaftliche Reihe, Nr. 4 (1964), 553.

the direct authority of the National Economic Council until the end of 1965, when it was divided up into individual industrial ministries. From 1966 these institutes were subordinated to their respective industrial ministry. Most of the other institutes in Table 3.1 were administered by the associations of state enterprises, the "VVBs".[1]

The "preponderant proportion of manpower concerned with natural scientific-technological R&D" in the GDR was located in the "centrally managed" sector of industry.[2] This sector contained enterprises grouped into VVBs, which in turn were directed by the National Economic Council and later, by the industrial ministries. The "regionally managed sector", by contrast, consisted primarily of small enterprises under the direction of regional economic councils. In 1964 this sector contained 86 % of all enterprises but its share of gross industrial production was only 28 %, and that of total employees, 29 %; the three main branches were food, wood, and clothing.[3] There were 1,095 research establishments in the centrally managed sector, and these incorporated 65.1 % of total R&D manpower.[4] Various publications put the total R&D manpower in 1964 at about 100,000[5] and this figure apparently excluded all or most of the social sciences and humanities.[6] The number of persons performing R&D in the centrally managed sector in 1964 was thus approximately 65,100, and the average manpower per establishment about 60.

Of interest is how this manpower was distributed between enterprise establishments and institutes, and what the average size of the two kinds of institution was. Table 3.2 shows that in the years 1961-1964 there were between 91 and 94 institutes having a total of 20,000 to 26,000 employees. The average manpower per institute was therefore around 250, a figure considerably higher than the overall average manpower per R&D establishment. Its order of magnitude compares quite well with the scattered data on individual institutes collected together in Table 3.3. All the figures relating to the sixties are in fact higher than the estimated 250, but the table has been compiled from information on the institutes published in journals and it could be the case that these institutes were rather more prestigious and larger than some of the others. To estimate the average size of enterprise R&D establishments, it will be assumed that all the industrial institutes belonged to the "centrally managed" sector of industry. Although some institutes conceivably conducted research association type work for enterprises in the "regionally managed" sector, they were almost certainly administered by the VVBs. Subtracting the number of institutes from the total number of R&D establishments in the centrally managed sector gives the figure 1,004 for the number of enterprise R&D establishments in 1964; a tenable figure given that there were 1,921 enterprises in the centrally managed sector in that year.[7] The estimated R&D manpower per enterprise establishment (excluding the 12 enterprise STCs) works out at 39.[8]

To summarise, these estimates suggest that the average size of the institutes in 1964 was around 250, that of enterprise R&D establishments around 40, and the overall average about 60. Comments and criticisms made in the literature about the size of R&D establishments give a certain credibility to these figures. An investigation of the R&D establishments in nine VVBs, which was probably carried out in 1963-1964, found that "small and the smallest"

Table 3.2
Total Number of Research Institutes and Personnel

	1961	1962	1963	1964
Number of Institutes	About 90[a]	94[b]	91[d]	
Manpower		About 20,000[c]	About 20,000[e]	About 26,000[f]

Sources:

a) H. Emmerich, "Zur Methodik des Planes 'Neue Technik' 1962", Sozialistische Planwirtschaft, Nr. 7 (1962), 19.
b) "Reserven aufdecken und nutzen", Die Wirtschaft, Nr. 20 (1962), 5; H. Such, VVB und wissenschaftlich-technischer Fortschritt (Leipzig: Staatsverlag der Deutschen Demokratischen Republik, 1964), 53.
c) "Reserven aufdecken...", op. cit.
d) K. Heuer, "Die wirtschaftliche Rechnungsführung in den naturwissenschaftlichen Instituten der sozialistischen Industrie und die Gewährleistung einer neuen Qualität der Planung, Leitung, Organisation und Wirtschaftsführung im wissenschaftlich-technischen Zentrum der VVB", Wissenschaftliche Zeitschrift der Humboldt-Universität zu Berlin, Mathematisch-Naturwissenschaftliche Reihe, Nr. 4 (1964), 553. NB: Heuer does not include three "scientific-technical bureaux" mentioned by sources b).
e) Ibid.
f) Such, op. cit., 127.

establishments were characteristic: 70 % had fewer than 36 employees, and 33.6 % had fewer than 16 employees.[9] In 1968 it was reported that 83 % of all R&D establishments employed fewer than 100 persons, 53 % fewer than 10, and many employed only one or two. The average size of the R&D establishments was given as 50.[10] Although these figures were probably not very different from those for many Western countries, GDR commentators, particularly in the period 1968-1971, felt they reflected a "fragmentation of effort" and an inefficient deployment of resources. Such views will be discussed more fully in Chapter Six.

Organisationally, the GDR in the mid-sixties appears to have had better conditions for technological change than the USSR. Soviet R&D was mainly centralised in facilities geographically distant from the place of production.[11] By contrast, about 60 % of the GDR's

Table 3.3
Manpower in Some of the Institutes
(Various Years)

	Manpower	Year	Source
Central Institute for Foundry Practice,	137	1953	a
Leipzig	300	1963	a
Central Institute for Production	81	1956	b
Engineering, Karl-Marx-Stadt	700	1968	b
Central Institute for Welding Practice,	80	1950	c
Halle	400	1962	c
Institute for Machine Tools,	166	1955	d
Karl-Marx-Stadt	460	1960	d
Institute Prüffeld for Electrical Heavy Duty Technology	300	1964	e
Institute for Fuel, Freiberg	583	1966	f
Average Manpower of the Institutes/STCs	250	1962-1964	g

Sources:

a) F. Naumann, "10 Jahre Institutsarbeit im Dienste der Gießereiindustrie", *Die Technik*, Nr. 7 (1963), 456.
b) "ZIF des Maschinenbaus", *Freiheit*, Nr. 49 (26. Februar 1968), 3.
c) W. Anders, "10 Jahre Zentralinstitut für Schweißtechnik der DDR", *Die Technik*, Nr. 5 (1962), 353.
d) K. Gläser, "Das Institut für Werkzeugmaschinen Karl-Marx-Stadt", *Die Technik*, Nr. 10 (1961), 683.
e) K. Heuer, "Die wirtschaftliche Rechnungsführung in den naturwissenschaftlich-technischen Instituten der sozialistischen Industrie und die Gewährleistung einer neuen Qualität der Planung, Leitung, Organisation und Wirtschaftsführung im wissenschaftlich-technischen Zentrum der VVB", *Wissenschaftliche Zeitschrift der Humboldt-Universität zu Berlin*, Mathematisch-Naturwissenschaftliche Reihe, Nr. 4 (1964), 553.
f) "10 Jahre Deutsches Brennstoffinstitut", *Neues Deutschland* (5. Oktober 1966), 2.
g) See text.

industrial R&D manpower was based in the enterprise and the remaining 40 %, that employed by the institutes, was also based for the most part near the place of production.[12] Furthermore the Soviet problem of administrative separation within the innovation cycle was probably not so acute in the GDR. To combat such administrative separation, the Soviet economic reform of 1965 gave the ministries authority over research institutes and enterprises. Nevertheless, the lines of command within the ministry remained separate, and the administrators were unlikely to have possessed much direct knowledge of enterprise needs and capabilities.[13] This system resembled the GDR's pre-reform centralised system of administering science and production, and GDR institutes were often criticised for conducting work of insufficient relevance to production and for indulging in "hobby research".

> The unsatisfactory association of many scientific research institutes with the burning technical-economic problems of the national economy's development ... led to the main tasks set by the Central Committee and the Council of Ministers not being undertaken and solved with the necessary intensity .... We have not been sufficiently successful in overcoming the dissipation of research and development work and the continual side-tracking onto secondary questions having no immediate bearing on the plan. There have been tendencies not to place science in the direct service of developing our material-technological base, but to pursue individual aims in research.[14]

The GDR's economic reform of 1963 enhanced the status of the VVBs, making them responsible for their branch of industry and in particular for the promotion and implementation of technological change.[15] The VVB, comprising headquarters, R&D institute, enterprises and enterprise R&D establishments, may be compared formally with a Western firm. The American scholar Rubenstein investigated R&D in large decentralised U.S. firms in 1960 and differentiated between three "pure" organisational forms: "centralisation", where R&D was conducted almost entirely in central laboratories; "decentralisation", where all R&D was conducted in divisional laboratories; and "combination", where R&D was conducted in both central and divisional laboratories. He found the "combination" model to be the most frequent; only in the raw materials and food sectors was the "centralisation" model of importance.[16] The GDR system had most in common with Rubenstein's "combination" model, whereas the Soviet system resembled his "centralisation" model.

Scherzinger, a West German, has asserted that many obstacles to transferring scientific results into industrial use stem from the much stronger separation of research and production in the GDR as compared with the FRG, and writes:

> For the Federal Republic it is known that in 1971 only about 5 % of the total research and development expenditure of industry was spent on research and development external to the enterprise ("betriebsextern"). A rough estimate shows that in the GDR more than a third of the research and development - measured in terms of the number of employees - is concentrated in supra-enterprise ("überbetrieblich") establishments of the national economy.[17]

This comparison does not support her assertion because the FRG statistic refers in fact to the "Unternehmen" (translated in this book as "firm") and not to the "Betrieb" ("enterprise").[18] In West Germany the terms "Unternehmen" and "Betrieb" are often used colloquially as synonyms, and economic literature sometimes takes the one and sometimes the other as the wider concept. Nevertheless official and semi-official statistics define the latter to be a geographically separate part of the former. A West German firm may therefore consist of a number of enterprises together with central, i.e. "supra-enterprise" R&D facilities.

Although industrial R&D in the GDR was organisationally more similar to that in the West than that in the USSR, in practice, as will be shown in this and later sections, it also shared some of the deficiencies of the Soviet system. One such deficiency was the problem of coordinating different VVBs for technological change of inter-branch significance. Another was the difficulty of coordinating institute and enterprise: after 1963 both, as a rule, came under the same VVB management, but there were still criticisms about the lack of relevance of some scientific work and about the pursuit of "hobby research". Yet it is unlikely that this problem was as severe as prior to the reform; VVB management had a better overview of their responsibilities than the National Economic Council, and financial measures were adopted to link research and production more closely together. An example of a VVB breaking the insularity of an R&D institute to outside developments and needs was given by Pöschel, advisor to the Central Committee on questions of research and development. In one important institute, some engineers had been developing a piece of equipment for eight years, even though the technical characteristics of the equipment were out of date at the beginning of the project. Although other members of the institute had pointed out the futility of the project, neither the institute director nor the party management subjected the matter to a critical examination. Only when the VVB intervened was the project stopped.

> For years on end, the manager of the development collective in question had argued that more modern pieces of equipment had not yet proven their operational ability in the national economy of the German Democratic Republic. Even when this "argument" had been refuted by facts, work on the project still continued, and the engineers tried to prove that for even such an out-dated piece of apparatus, use could surely be found somewhere in our national economy.[19]

Between 1968 and 1971 the organisation of research, development and production in the GDR underwent extensive reform, with enterprises and research facilities being merged into "combines". One director was given overall responsibility for the combine and he had to account for its performance either to the VVB headquarters or direct to the ministry. Unlike the VVB arrangement, the combine could be a vertical integration of enterprises, thus encompassing the whole or much of the "science-production-cycle". Furthermore, various managerial techniques such as "task management" and "transfer teams" were introduced to help circumvent discontinuities in the process of technological change. Some similar reforms also took place in the Soviet Union from about 1970, albeit on a relatively smaller scale.[20] The GDR's reorganisation will be considered more fully in Chapter Six.

FACILITIES FOR INDUSTRIAL R&D

Large and "development-intensive" industrial enterprises usually had a self-contained R&D department or sphere, and smaller enterprises accommodated R&D work in the enterprise design office or laboratory.[21] As in the Soviet Union, there were problems concerning the supply of R&D materials, the provision of scientific equipment, and the capacity for "development-transfer" work.[22]

Günter Mittag, the leading functionary for industry and candidate member of the Politbureau, stated in 1966 that there were long delivery periods for even the smallest quantities of R&D materials. This, he said, meant that the "usual international deadlines" in research work could not be kept, and caused "unproductive expenditure of labour".[23] Prestigious research establishments, particularly if conducting work considered important by the central authorities, were probably in the most fortunate position to obtain supplies, although sometimes the intervention of an industrial minister was needed to secure components.

> The Central Institute for Production Engineering needed literally two years for the further automation of the box column drill shown at the "Technica" (Fair), a time in which a medium-sized investment project which includes construction work can normally be completed. And if "Technica 66" had not been imminent, and a missing coupling had not been procured from VEB Elmo Dessau on the energetic directive of the minister responsible, the Central Institute for Production Engineering would probably still not be finished with the special purpose modernisation of the box column drill.[24]

The lower level of GDR technology relative to the FRG particularly, as suggested in Chapter Two, in branches like scientific and electrical measuring instruments, electronics, and computers implies that industrial R&D facilities in the GDR were, as a rule, considerably less well equipped than those in the FRG. The first official survey of industrial R&D facilities was carried out for the year 1965 and used the index "fixed assets per R&D worker". It indicated that facilities varied widely between establishments; the highest allocation of fixed assets (excluding buildings) per R&D worker being six to eleven times as great as the lowest allocation.[25] Meyer of the State Central Administration for Statistics questioned whether the low allocation in certain establishments was a reason for overlong R&D times. Further, he said that since fixed assets for R&D were to create the preconditions for the "technical revolution" and especially for competitive products and processes, the proportion of fixed assets older than ten years (between 6 % and 42 %) was high.[26] A later source stated that the index "fixed assets per industrial R&D worker" was higher in the seventies than it had been in 1965, though there were variations between industrial branches.[27] The economist Meske pointed out that in the chemical industry, the index was lower in 1970 than 1965 since the rate of equipping had not kept pace with the large increase in R&D personnel.[28]

It should be noted that important problems are involved in the use of this index as an indicator of R&D facilities as it can be inflated by factors such as insufficient discarding of old or no longer used fixed assets, by domestic price changes, and by increased prices for equipment from capitalist countries. It would therefore be unwise to conclude from the higher level of fixed assets per R&D worker in 1970 compared with 1965 that the increase of R&D manpower over this period was more than matched by an increase in R&D facilities: insufficient information is available to decide the point. What may be said is that in the late sixties and the seventies, increased efforts were made to improve the supply of R&D materials and the provision of scientific equipment. Scientific instrument building was classified in 1968 as a priority, growth branch; maximum use of apparatus was to be ensured by promoting joint utilisation and shared investments; and at the Ninth Party Congress in 1976, Honecker called for a strengthening of the scientific instrument building industry.[29]

Capacity for development and "transfer" work was criticised as inadequate. At the Seventh Party Congress in 1967, for example, all managers and directors were urged to pay more attention to expanding capacity for the transfer of research results into production:

> Many examples show that progressive scientific-technological results which had been achieved in the shortest time period remained unused because capacities in prototype and pilot lot construction, jig and tool-making, or in design and technological preparation were not sufficiently developed.[30]

As in the Soviet Union, available facilities were not always used for development work. An investigation of twenty-one VVBs published in 1964 found that 65 % of the existing capacities for prototype and testing model construction were utilised for current production: an estimated 1,500 highly qualified persons were thus working on tasks not directly related to R&D.[31] Similarly a source published in 1966 stated that in many enterprises, between 60 % and 70 % of "design and technological capacity" was deployed for current production.[32]

The planners attempted in at least two ways to ensure that new development-transfer facilities would not be requisitioned for current production plan fulfilment. The first method was initiated in the late fifties and involved the founding of "scientific-industrial enterprises" which conducted R&D work, made testing models, undertook design to the production stage and sometimes manufactured pilot lots.[33] Such enterprises differed from the usual type of enterprise in that R&D accounted for a large portion of the work, and production was mainly on an individual or small-series basis.[34] The second method involved the allocation of production facilities to industrial institutes. The technical director of "VVB Automobile Construction", Dr. Opitz, stated in 1966 that his VVB had established a testing and development department together with corresponding production capacity. This enabled the institute to transfer small series of developments into production, and the experience had been "extraordinarily positive".[35] Neither method appears, however, to have been used to any appreciable degree, and one must in any case be sceptical that they could fully escape the imperative of the VVB's production plan.

GDR publications of the seventies indicate that capacity for development-transfer work continued to be a problem, although this is not obvious from statistics on the proportion of expenditure devoted to basic research, applied research and development. Table 3.4 shows that the GDR had a similar distribution of expenditure to that in the industrialised capitalist countries whereas the Soviet Union had a clear imbalance between research and development.[36] In the GDR the main weaknesses were the continued use of development-transfer facilities for current production and, as will be seen in the section on the employment and qualification structure of industrial R&D manpower, a shortage of qualified personnel, especially "technologists", on the interface between development and production.

TIME SPENT ON R&D

The work performed by the institutes and enterprise R&D establishments included activities which would not fall within the OECD's "Frascati" definition of research and development.[37] This is necessary to bear in mind when attempting to compare the R&D effort of the GDR with that of Western countries. Günter Mittag reported in 1966 that development engineers of important large enterprises had less than 50 % of their working time available for actual research and development. "The remaining time is taken up with administrative work, writing reports, and subordinate activities."[38] Unfortunately, he did not specify these duties in any greater detail; the Frascati definition would certainly include administrative and report-writing work which was inherent to R&D activity. Other GDR sources, however, permit one to distinguish at least three major types of "non-Frascati" work performed by industrial R&D manpower.

The first kind of non-Frascati activity was conducted in the institutes. The Central Institute for Production Engineering, for example, was said in 1963 to perform not only oriented research and development, and contract work for other institutions, but also

Table 3.4
Distribution of Expenditure over Basic Research, Applied Research and Development in 1971 (%)

	Basic Research	Applied Research	Development
GDR (1972)	13.5	30.2	52.0
USSR (1968)	12.9	60.5	26.6
Belgium	24.6	33.3	42.1 (a)
Canada	21.4	38.0	40.6
Finland	17.5	32.5	49.9
France	18.5	35.0	46.6 (a)
FRG	26.9	73.1	
Greece	21.9	48.6	29.5 (a)
Iceland	24.0	59.6	10.5
Ireland	9.8	36.9	53.3
Italy	15.3	43.9	40.8 (a)
Norway	18.8	32.5	48.7
Spain	17.8	36.4	45.7
Sweden	16.7	19.8	63.5
UK (1968)	10.9	25.7	63.4 (a)
USA	14.7	22.6	62.7

Sources:

Autorenkollektiv, Das Forschungspotential im Sozialismus, (Berlin: Akademie Verlag, 1977), 88; OECD, Survey of the Resources Devoted to R&D by OECD Member Countries, International Statistical Year 1971, Vol. 5, Total Tables (Paris: OECD, 1973), 42-44; OECD, International Survey of the Resources Devoted to R&D in 1969 by OECD Member Countries, Vol. 5, Total Tables (Paris: OECD, 1973), 73.

Note:

(a) = Total intramural expenditure on R&D. The other figures for the capitalist countries give the current intramural expenditure.

tasks concerning information and documentation.[39] It is likely that the central institutes subordinate to the ministry spent a higher proportion of their working time on R&D than the VVB institutes. The latter were also concerned with questions of organisation and production, although there seems to have been considerable variation between them with regard to the type of work done.[40] There were also differences of opinion amongst GDR writers as to what role these institutes should have. Zander stated in 1966 that a VVB institute ought to be concerned with the following areas: documentation and information, market research, R&D, standardisation, design, organisation, technology, production, quality and economics.[41] Kerda, on the other hand, rejected any such formal stipulation of duties, and stressed that the function of a VVB institute should depend upon the specific conditions and needs of the particular industrial branch. Where enterprise R&D was weak, the institutes had to involve themselves in product development to design stage; otherwise they could concern themselves with more basic research questions of importance for the whole branch.[42] Economists Kusicka and Leupold noted that:

> In the institutes of several VVBs we investigated, the time spent on oriented basic research, standardisation tasks, contract research and effective support of enterprise production only amounted to between 20 and 50 percent of the total working time. And even this time was not fully used for creative scientific activity.[43]

The second non-Frascati activity characterised both institutes and enterprise R&D establishments and involved routine work in R&D itself. The general director of "VVB Components and Vacuum Technology", Heinze, stated in 1964 that as a result of poor organisation, development personnel had to spend up to 50 % of their time obtaining "the material prerequisites for their work".[44] According to Kusicka and Leupold, individual R&D workers had to spend between 20 % and 30 % of their time on routine activities.[45]

A third non-Frascati activity particularly prevalent in enterprise R&D establishments, but also to be found in VVB institutes, related to production. R&D departments were supposed to render "effective support of production", though this formula seems to have been given a liberal interpretation by many enterprise and VVB managements. Sometimes R&D personnel were called upon to deal with small technical problems which should have been solved by the enterprise technical staff, the "innovators" and the "worker researchers". On other occasions R&D personnel were, as noted in the previous section, delegated to help out with the production plan and other tasks. One such example was that of a well-known television enterprise which, in 1962, requested a similarly well-known institute for high frequency physics to conduct some important research on its behalf since its own development and design department could not solve the problem. After the work had been completed, the institute sent a report to the enterprise towards the end of 1961. There was no reaction to the report. A quarter of a year later, in February 1962, the institute management asked how the report had been received. A reply was sent from the enterprise, signed by the technical director and chief designer, stating: "As a result of the deployment of the whole development department in production for several months, it has not been possible for us to look through the work earlier."[46] According to one source, R&D workers were "enlisted for all kinds of jobs, often unconnected with their duties, on the assumption that manpower could sooner be spared from research, and for a few hours it did not matter".[47] In fact the R&D department tended to be regarded as a kind of "fire brigade" which could be called out when problems arose. In the early sixties, the practice was particularly reinforced by two important features of industrial R&D establishments. Firstly, small R&D establishments were frequently manned by poorly qualified personnel and were not capable of performing proper R&D work. Secondly, perhaps as a reaction against "hobby research", the performance of VVB institutes was sometimes judged on the basis of how often and for how long institute R&D personnel had visited the enterprise to help improve the organisation of labour and production. These features were possibly less pronounced in the seventies, although the practice of detailing R&D workers to assist with the production plan continued.

> It came to a heated discussion in the research and development department as the enterprise manager detailed ten workers from this section into production. The readiness to help out directly at the work bench at weekends and in extra shifts is considerable even for designers, engineers and scientists. Setting an example was even the manager of the main R&D section, Dr. Ing. Friedrich-Wilhelm Bretschneider, who worked a milling machine for several weekends. What is not understood is that during regular working hours, highly qualified manpower is assigned for such auxiliary tasks as painting or the removal of rust. Of course even the works management is aware that plan-backlogs cannot be made up by the deployment of R&D workers in production. But it thinks it can make a virtue out of necessity by justifying the decision on the basis that it would not do the R&D workers any harm to learn on the spot how the results of their work are implemented.[48]

EMPLOYMENT AND QUALIFICATION STRUCTURE OF INDUSTRIAL R&D MANPOWER

Table 3.5 gives a breakdown of the average employment structure of R&D establishments in the centrally managed sector of industry; the figures most likely refer to 1963-1964. Particularly noticeable is the high percentage of specialists not possessing university or technical college training. The proportion of persons employed in research, development and prototype construction was 64.3 % and according to Kusicka and Leupold it lay below that to be found in capitalist concerns, i.e. 70-75 %.[49] Such proportions, however, depend upon what is included for measurement. The category "other employees" in the table includes "administrative personnel, maintenance personnel, enterprise guard, drivers, cleaning staff, and others", many of whom provide an indirect service to R&D and as such would be excluded from the measurement of R&D manpower according to the "Frascati" definition. If the category "other employees" were reduced by a factor one-third, for example, the proportion of specialists engaged in research, development and prototype construction would become 73 %.

To assess the significance of the figures in Table 3.5, an experimental comparison with the FRG has been made. Data on R&D manpower in West German firms are available for the year 1965, and the results are shown in Table 3.6. The comparisons suggest that the GDR had approximately the same proportion of

Table 3.5
Average Employment Structure of R&D Establishments in the Centrally Managed Sector of Industry

	%	
Specialists in Research and Development	44.7	
of which: University Graduates		11.1
Technical College Graduates		23.3
Non-Graduates		10.3
Employees in Prototype Construction	19.6	
Other Employees	35.7	
Total	100.0	

Source: Based on H. Kusicka and W. Leupold, *Industrieforschung und Ökonomie* (Berlin: Dietz Verlag, 1966), 44.

Table 3.6
Experimental Comparison of Employment and Qualification in Industrial R&D for the Two Germanies

	GDR (1963/64) %	FRG (1965) %
R&D Specialists		
of which: QSEs	12.6	12.7
Qualified Technicians (QTs)	26.4	42.8
Unqualified Specialists	11.7	-
Other Supporting Staff	49.2	44.5
Total	99.9	100

Source:

Based on Table 3.5 and H. Echterhoff-Severitt, "Wissenschaftsaufwendungen in der Bundesrepublik Deutschland. Folge 2: FuE-Personal in Unternehmen und Verbänden", Wirtschaft und Wissenschaft, Nr. 2 (1969), 22.
For details of the methodology see note 50.

qualified scientists and engineers (QSEs) in industrial R&D as West Germany; a somewhat higher percentage of "other supporting staff"; but a low ratio of qualified technicians (QTs). This inference is also supported by other statistical data for the two Germanies. Table 3.7 gives the ratio QSE to QT for R&D in three industrial sectors of the GDR, and again rough comparison with the FRG (Table 3.8) suggests a low proportion of qualified technicians in the GDR. This situation appears to have persisted into the seventies because a "representative questionnaire" conducted amongst R&D workers revealed that "the main factor" for deficiencies in the research process was the lack of technical personnel.[51]

Although the overall proportion of qualified scientists and engineers to the remaining staff in the mid-sixties was roughly the same in the GDR as the FRG, there are indications that there was a shortage of QSEs in some enterprise establishments. Complaints were made by the leadership about the lack of graduates in the enterprise, and especially the low number of scientific-technical graduates. Apparently a "proportion" of the enterprise R&D establishments had only a small personnel, and many of these workers had low qualifications.[52] An investigation of nine VVBs revealed that 18.6 % of all R&D establishments had no university graduates on the staff; 92.9 %

Table 3.7
Ratio QSE to QT for R&D in Three Industrial Sectors of the GDR (1966)

Extraction Sector	1 : 1
Processing Sector	1 : 1
Branch of Instruments of Production	1 : 1.5

Source:

Figures taken from W. Wolter, "Zur Ökonomie der Hochschulbildung", in A. Knauer, H. Maier and W. Wolter (eds.), Bildungsökonomie. Aufgaben - Probleme - Lösungen (Berlin: Verlag Die Wirtschaft, 1968), 144.

Notes:

(1) The editing of this book was completed on 30.10.1967, so it is assumed the statistics refer to 1966.
(2) Values are rounded.

Table 3.8
Ratio QSE to QT for R&D in Three Industrial Sectors of the FRG (1965)

Mining Sector	1 : 2.2
Manufacturing Sector	1 : 3.4
Steel, Machine and Vehicle Construction Sector	1 : 4.4

Source:

Calculated from H. Echterhoff-Severitt, "Wissenschaftsaufwendungen in der Bundesrepublik Deutschland. Folge 2: FuE-Personal in Unternehmen und Verbänden", Wirtschaft und Wissenschaft, Nr. 2 (1969), 22.

had fewer than eleven; and 99.3 %, fewer than forty-one (Table 3.9).[53] In the USA, reportedly 68 % of the industrial R&D establishments had fewer than eleven graduates; and 90 % fewer than forty-one; one percent had more than 400.[54]

Too few "technologists" (Technologen), especially highly qualified "technologists", were employed in the industrial enterprises. The specific duties of the "technologist" varied from industrial branch to branch, though they most likely included work on the development-production interface.[55] The ratio of production workers to "technologists" in the branch

Table 3.9
University Graduates on the Staff of R&D Establishments (1963-1964)

% of R&D Establishments	Number of University Graduates From	To
18.6	-	-
67.1	1	5
7.2	6	10
2.8	11	20
3.6	21	40
0.7	41	100
100.0		

Source: H. Kusicka and W. Leupold, Industrieforschung und Ökonomie (Berlin: Dietz Verlag, 1966), 53.

"automatic control technology" was given in 1967 as 16.5 : 1. The ratio of production workers to graduate "technologists" was 205 : 1. Bernicke stated that in the USSR there were ten production workers to one "technologist", and a considerably higher proportion of "technologists" possessed university qualifications than in the GDR.[56] Fritzsche pointed out in 1968 that persons in the "technology" departments generally had lower qualifications than, for example, in design; the situation was not solely to be remedied by raising salaries, it was also important to reduce the large amount of routine work.[57]

As in the Soviet Union then, the most highly qualified personnel and largest concentration of R&D manpower were to be found in the institutes. The shortage of qualified scientists, engineers and "technologists" at the enterprise level together with the problem of development-transfer facilities and the use of manpower for non-R&D work suggest that enterprise R&D was frequently weak. In the remainder of this section, some related reasons for the shortage of qualified personnel in the enterprise will be discussed.

The effect of the Second World War on the population structure and the migration of an estimated one-third of the GDR's academics to West Germany prior to 1964[58] meant that in principle there was a need for qualified persons at all levels of the economy. Status considerations played a decisive role in job selection: highly qualified scientists and engineers preferred ositions in the Academy of Sciences or the universi-ies to ones in industry; within industry, institutes ere more prestigious than enterprise establishments, nd R&D departments more highly rated than "techno-ogy" departments. Evidently a number of people managed o obtain posts in the enterprise R&D establishments ven though their qualifications were not commensurate ith the level of work to be done; once there they tayed, although later they could not keep pace with ncreasing requirements.

Another reason why there were too few qualified cientists, engineers and "technologists" at enterprise evel was that the central authorities were not con-picuously adept at allocating R&D manpower. Walter lbricht complained in 1963 about the work of the ational Economic Council as follows:

> How, for example, do the comrades in the Section for Ferrous Metallurgy of the National Economic Council conceive the development of their industrial branch when they only want to appoint five mathematicians and three physicists up to 1965? According to the notions of the National Economic Council sections, only about 30 % of the graduates in mathematics, physics and chemistry can allegedly be employed in the coming year. These comrades have apparently not understood what is at stake in the economic struggle between socialism and capitalism.[59]

n 1964 the council allocated 91 natural scientists o the "VVB Components and Vacuum Technology". The ontingent included seven medical electrologists, hree nuclear physicists, and a radio chemist: spe-ialists whose training the VVB could not properly se.[60] Similar situations also occurred in other VVBs.

A third reason for the shortage of QSEs at enter-rise level was that enterprise and VVB management ften had little interest in employing them.

> Many economic functionaries give important lectures on the role of modern natural science in the development of socialist production. But they themselves are not yet clear about the deep importance of this problem, as otherwise they would try their utmost to engage mathematicians, physicists, chemists etc., and to win as large a number as possible for their enterprise. Instead, the VVBs and enterprises this year had to be issued with administrative directives to take on graduates.[61]

This lack of interest was also manifested in the assignment of R&D manpower to routine or minor enterprise tasks and was reflected in the lack of clear ideas about future requirements of QSEs. A report published in 1964 stated that a "recommendation" from the Bureau for Industry and Construction (Politbureau level) to the leading functionaries of a number of VVBs was first necessary before they concerned themselves with the likely needs of scientific-technical manpower, and even then only very general conceptions were passed on.[62] This created problems for those concerned with planning the output of graduates from the universities. A publication which appeared in 1967 pointed out that if previous estimates from industry had been accepted, mathematics education would have been cut down considerably; currently, owing to the development of electronic data processing etc., there was no hope of meeting the demand for mathematicians.[63] In some cases management seems even to have been positively opposed to employing graduate specialists. "One management comrade from the enterprise 'Electrocoal' said in this connection that they had regular bouts of anxiety when it came to placing graduates in the enterprise."[64]

The disinclination or aversion to employing scientists and engineers in the enterprise had both financial and psychological aspects. Enterprises, as will be shown in Chapters Four and Five, had inadequate financial incentives to pursue technological change actively. Moreover, productivity indicators could work in a peculiar fashion: fewer R&D workers in the enterprise meant that the index "labour productivity" increased and that the relationship between "productive" and "unproductive" labour "improved".

> The enterprise Secura Berlin, for instance, which until recently had severely underestimated research and development work anyway, received in 1962 the instruction to reduce the research and development section by 11 posts in order to correct the "disproportion" between the production workers and the "remaining personnel". After it became evident, however, that great sales difficulties compared with competitors' products were to be expected on the capitalist market, the research and development section had to be staffed more strongly again. Similarly the VEB Factory for Television Electronics Berlin had its staff appointments plan for research and development reduced by 25 posts in 1963 by the VVB Components and Vacuum Technology, even though

> the enterprise has to solve important problems in the electronics field. There has been a good deal of such nonsense.[65]

Important managerial posts were filled in the period of postwar reconstruction by convinced and proven socialists, who as a rule neither belonged to the scientific-technological intelligentsia nor possessed high professional competence.[66] They, in common with many party members, regarded bourgeois specialists with ideological suspicion. When intellectuals and professionals were granted a number of privileges in the late forties and early fifties to deter further migration, the Politbureau felt it necessary to criticise those comrades who still paid homage to "levelling" and who disregarded the achievements of the intelligentsia.[67] The fundamental concern for most managers was probably to obtain or retain subordinates who were not likely to endanger their personal power positions. Such psychological mechanisms are, of course, a common feature of all organisations, socialist and capitalist, and left unchecked are both self-perpetuating and stifling to productivity. Enterprises were often proud if they could reconstruct their facilities without having to employ highly qualified professionals. National Prize Winner Professor Frühauf gave examples for the electrical industry in 1954: one enterprise with about 2,000 employees had no diploma-engineers; another with around 1,000 employees had only one, and the management of this enterprise was of the opinion that it did not need any newly trained engineers.[68] Such attitudes tended to persist into the sixties: there was a widespread opinion in industrial management that skilled workmen and foremen were, as a result of their long experience, in a better position to help fulfil the enterprise plan than young graduates. Even when graduates were appointed in the enterprises, they were frequently employed in work not corresponding to the level of their training. The works manager of the combine "Otto Grotewohl" actually proposed an "economic experiment" in which all graduates of universities and technical colleges be first employed as skilled workers until they had acquired enough knowledge to be promoted to the next higher position.[69]

By the sixties, however, political reliability was no longer a sufficient justification for holding a top post in industry. What the New Economic System required was managers with professional ability, and there were heightened fears that if young graduates were engaged, they could advance quickly and threaten the positions of those managers who could no longer cope with their increased responsibilities. According to Ulbricht, managers would be of little use in the New Economic System who were "petty, narrow-minded red-tapists" who, in isolation from life and the struggle of the working class, had lost both an appreciation of newness and contact with people. Just as little use would be "old hands" who did not want to understand that it was impossible to manage personnel with old accustomed, administrative and dogmatic methods.[70]

During the NES and early seventies, the employment situation in industry of graduates and particularly of qualified scientists, engineers and "technologists" improved: between 1961 and 1971 the number of graduates working in industry per 1,000 industrial employees almost tripled.[71] Job opportunities in the academic sector for qualified scientists and engineers contracted, and the older generation of industrial management increasingly moved into retirement. Even so there were still reports in the seventies about the lack of graduates in industry, in R&D, and especially in "technology". The latter continued to have a poor status: one observer described it as a "handmaiden of research".[72]

> Connected with this have been practices which do not give proper recognition to the achievements of the technologist compared with the research and development worker, and which separate product and process development, design and the technological preparation of production from each other. For these reasons some engineers regard their employment in the technology department as "beneath their dignity". All this is a result of the long-standing underestimation of technology, which, in my opinion, stems all the way back to educational training.[73]

THE SCALE AND DIRECTION OF THE INDUSTRIAL R&D EFFORT

Fragmentary data on the total number of persons engaged in "scientific-technological R&D" in the GDR between 1963/64 and 1975 are assembled in Table A.1 of Appendix A.[74] The figure cited in the literature for the sixties was 90,000 - 100,000 persons, and that for the first half of the seventies, 150,000 - 155,000. It is not clear, however, precisely what activities constitute "scientific-technological R&D". If the figures were accepted at their face value, the GDR would appear to have had a lower absolute R&D effort than the FRG, but a higher R&D effort in per capita terms. Table A.2 in the Appendix presents a formal comparison of total R&D manpower in the two Germanies for the years 1963/64 and 1973. In both these years the total absolute R&D manpower in the GDR was roughly 50 % the size of that in the FRG. On the other hand, the population of the GDR was less than one-third that of the FRG. A similar impression would be gained from a formal comparison of the number of persons engaged in industrial R&D in the two Germanies. GDR data are available for the years 1963/64 and 1970 and, as may be seen from Table A.3, the GDR's industrial R&D effort also appears to have been about 50 % the size of that in West Germany.

Yet as noted in foregoing sections, GDR figures are inflated when considered in relation to the OECD's Frascati definition as they include persons providing an indirect service to R&D, and R&D manpower performing duties that were not R&D in the narrow sense. Appendix B contains an experimental attempt to compare the industrial R&D effort in the two Germanies according to the Frascati definition. The conclusion is that in both years for which data were available (i.e. 1964 and 1970), the GDR's industrial R&D effort was not only lower than that of the FRG in absolute terms but was most likely lower in relative terms also. The measure adopted for the calculation was the ratio of qualified scientists, engineers and technicians to industrial employees.

The FRG is as rigorous a yardstick in its levels of R&D as it is in its levels of technology, and it is worthwhile attempting at least some rough comparisons of the GDR with other OECD countries. Table A.4 contains statistics on the industrial R&D manpower of various countries in 1963/64. It shows that the FRG came second in absolute terms after Japan, and first measured relative to one million population. Nevertheless even if GDR figures were deflated by 50 % to adjust for non-Frascati activities, the GDR would still compare very favourably with the other countries. Its absolute industrial R&D manpower would be 32,550 persons and it would come fourth in the list, after Japan, the FRG, and France, but before Italy. Per million population the GDR would have 1,897 persons engaged in industrial R&D and would come third in the list after the FRG and Sweden, but before Japan. Another rough comparison is contained in Table A.5 which shows the proportion of industrial R&D manpower to total scientific R&D manpower for various countries. In 1971 about 55 % of the GDR's R&D manpower worked in industry, which was lower than the proportion for the FRG (69 %), Belgium (62 %), and Japan (60 %) though comparable to that for Sweden (58 %), France

(54 %) and Italy (53 %). This comparison is also a tentative one since certain contents of the OECD categories "private non-profit" and "government" may possibly fall under the GDR's definition of "industry".

A more detailed idea of the size and direction of the GDR's effort can be gained by examining statistics relating to the various branches of industry. The distribution of R&D manpower by type of industry in 1963 is given in Table A.6 and the figures are compared with statistics for the Federal Republic in Table 3.10. This table suggests two main differences between the research efforts of the two Germanies in the early to mid-sixties: there was a much greater emphasis in the GDR on machine building than in the FRG (51 % and 21 % respectively), and the GDR had a considerably smaller proportion of R&D manpower in the chemicals sector (14 % and 34 % respectively). Again owing to differences in definitions, Table 3.10 can only be a rough comparison: the GDR figure for "machine building" is probably inflated and contains items that would come under other headings in the FRG taxonomy, a likely example being "fine mechanics and optics" which in the FRG was classified under "electrotechnical and electronics".

The proportion of total industrial R&D effort devoted to chemicals in the GDR was low compared not only with the FRG but most other OECD countries. Belgium had the figure 40 %; Italy, 38 %, the Netherlands, 31 %; Japan, 28 %; Austria, 23 %; Canada, 22 %, and France, 20 %.[75] The United States and Britain had similar proportions to the GDR but these countries devoted about one third of their total industrial R&D effort to aircraft.[76] Given that the GDR, like the FRG, inherited a highly developed chemical industry at the end of the war, that chemicals have accounted for a sizeable proportion of total GDR production and exports,[77] and that the country introduced a "Chemical Programme" in 1958, it would however be surprising if absolute and per capita figures for chemical R&D manpower in the GDR were extremely low. Table A.7 shows in fact that even if the GDR figures were deflated by 50 % to compensate for non-Frascati activities, the GDR would be roughly on a level with Belgium in absolute terms, and with France and the Netherlands measured relative to one million population.

Other available statistics suggest that the GDR's R&D effort was particularly weak relative to that of the FRG in modern growth areas. Table A.8 contains data on the ratio of R&D manpower to work force for several VVBs in 1964, and these are compared with figures for the industrial sectors of West Germany

Table 3.10
Distribution of R&D Personnel Over Industrial Sectors of the GDR (1963) and FRG (1965)

Sector	GDR	FRG
Energy and Mining	2.5[a]	2.1
Chemicals	13.7	33.9[c]
Ceramics and Glass	2.3	0.9
Ferrous and Non-Ferrous Metals	4.0[b]	7.3
Machine Building	51.0	21.5
Electrotechnical/Electronics	22.7	29.0[d]
Wood, Paper, Printing	0.9	0.4
Textiles, Clothing, Leather	1.2	2.0
Food	1.6	0.5
Others	0.1	2.4
Total	100	100

Sources:

GDR figures from Table A.6. FRG figures based on H. Echterhoff-Severitt, "Wissenschaftsaufwendungen in der Bundesrepublik Deutschland, Folge 2: FuE-Personal in Unternehmen und Verbänden", *Wirtschaft und Wissenschaft*, Nr. 2 (1969), tables 8 and 9 on pages 22 and 23; Folge 5: "FuE-Personal in Unternehmen und Verbänden im Jahre 1969", *Wirtschaft und Wissenschaft*, Nr. 6 (1971), table 17 on page XIX. Estimate for association R&D in chemicals and leather/textiles based on 1969 figures.

Notes:

a) Includes State Geology Commission.
b) Includes Potash.
c) Includes Synthetic Products and Rubber.
d) Includes Fine Mechanics and Optics.

in Table 3.11. Since the GDR figures have not been adjusted to the Frascati definition they are slightly inflated. Whereas VVBs in machine building (with the possible exception of that for textile machines) and some in the electrotechnical sector compared well with the FRG, comparatively low ratios were held by VVBs in chemicals, plastics processing, data processing, and electrical machine and apparatus construction. Table A.9 gives statistics on the ratio of university graduates to total R&D manpower for a few VVBs and research institutes in 1964. Although the ratio for the centrally managed industry as a

Table 3.11
Ratio of R&D Workers to Total Employed in 1964/65 (%)

GDR (VVB) in 1964		FRG (Industrial Sector) in 1965	
Synthetic Fibres and Photochemicals	1.94	Synthetic Materials and Rubber	2.30
Electrochemicals and Plastics	2.11		
Rubber and Asbestos	1.70		
Plastics Processing	0.98		
General Chemicals	2.70	Chemicals and Mineral Oil	7.89
Mineral Oils and Organic Basic Materials	2.79		
Pharmaceutical Industry	3.78		
Paints and Varnish	2.44		
Mining Equipment and Conveyance Plant	4.40	Steel, Machine and Vehicle Building	3.84
Chemical Plant	4.66		
Textile Machine Building	2.73		
Polygraph, Machines for Paper and Printing	3.68		
Food and Packing Machines	3.75		
Machine Tools	3.54		
Data Processing and Office Machines	2.73	Electrotechnical, Fine Mechanics, Optics	5.71
High Tension Equipment and Cable	3.42		
Components and Vacuum Technology	3.86		
Automatic Control Equipment Construction and Optics (excl. VEB Carl Zeiß)	4.96		
Communications and Measurement Technology	6.80		
Electrical Machine Construction	3.24		
Electrical Apparatus Construction	2.50		
VEB Carl Zeiß Jena	6.70		

Source:

Based on Table A.8 in Appendix A and H. Echterhoff-Severitt, "Wissenschaftsaufwendungen in der Bundesrepublik Deutschland. Folge 2: FuE-Personal in Unternehmen und Verbänden", Wirtschaft und Wissenschaft, Nr. 2 (1969), 22.

whole (11.1 %) was roughly the same as that for West Germany,[78] low values were held by the institutes for automatic control technology and conveying technology and by the VVB for data processing and office machines (6.9 %). In the FRG the ratio of QSEs to total R&D manpower in the category "instruments, other electrical machines and apparatus" for 1964 was 15 %.[79]

Thus one reason for the large inter-German technology gap in "modern" branches such as plastics, electronics, data processing, and instrument building was presumably a substantially lower GDR research and development effort in these branches, in both absolute and relative terms. It is possible that this situation in R&D had improved by the seventies. Although no detailed statistics on the GDR's R&D by industrial branch are available for these years, the rough estimate contained in Table A.10 suggests a considerable increase over 1963 in the shares of total industrial R&D manpower employed in the electrotechnical, electronics and instrument building sector and in the chemical sector. These figures must be viewed with caution, however, because of the method of their derivation and the fact that the GDR changed its statistical definitions between the sixties and the seventies.[80] What can be said is that the "modern" branches were identified by the party leadership as being vital for the "technical revolution", they experienced higher than average growth rates in the sixties and, as will be shown in Chapter Five, received priority promotion under the "offensive strategy for technological change".

> We speak of technical revolution because whole industrial sectors experience a radical change; because with the help of electronic data processing units the management of the economy, the control of production processes, and administrative activities are reorganised and made much more effective; because automation, the application of cybernetics, chemicalisation and the utilisation of nuclear energy ever increasingly determine modern production.[81]
>
> Walter Ulbricht

NOTES

1. H. Such, VVB und wissenschaftlich-technischer Fortschritt (Leipzig: Staatsverlag der Deutschen Demokratischen Republik, 1964), 53 and 131; Autorenkollektiv, Die Finanzen der Industrie in der Deutschen Demokratischen Republik (Berlin: Verlag Die Wirtschaft, 1966), 177.

2. H. Kusicka and W. Leupold, Industrieforschung und Ökonomie (Berlin: Dietz Verlag, 1966), 38.

3. Calculated from St.J.DDR 1965, 106-109. NB: The cooperative, half-state and private enterprises have been grouped here under "regionally managed industry".

4. Kusicka and Leupold, op. cit., note 2.

5. A. Springer, "Die Effektivität der wissenschaftlichen Forschung erhöhen", Einheit, Nr. 6 (1966), 818; H. Pöschel, Leitung von Forschung und Technik und wissenschaftlicher Meinungsstreit (Berlin: Dietz Verlag, 1964), 51; H. Such, VVB und wissenschaftlich-technischer Fortschritt (Berlin: Staatsverlag der Deutschen Demokratischen Republik, 1964), 127; Kusicka and Leupold, op. cit., note 2, 43. See also Table A.1 in Appendix A.

6. See W. Meske, "Zur Entwicklung des Forschungspotentials der DDR unter den Bedingungen der intensiv erweiterten Reproduktion im Sozialismus", in Sozialistische Wissenschaftspolitik und marxistisch-leninistische Wissenschaftstheorie, Kolloquien Reihe des Instituts für Wissenschaftstheorie und -organisation an der Akademie der Wissenschaften der DDR, Heft 10 (1975), 116.

7. St.J.DDR 1965, 106.

8. It is assumed there were 91 institutes in 1964 having a total of 26,000 persons.

9. Kusicka and Leupold, op. cit., note 2, 52.

10. D. Graichen and L. Rouscik, Zur sozialistischen Wirtschaftsorganisation. Aufgaben - Probleme - Lösungen (Berlin: Verlag Die Wirtschaft, 1971), 225.

11. R. Amann, J. Berry and R.W. Davies, "Science and Industry in the USSR", in E. Zaleski et al., Science Policy in the USSR (Paris: OECD, 1969), 403-404.

12. This assumes that the number of persons employed in the institutes was 26,000.

13. Amann, Berry and Davies, op. cit., note 11, 404, 432-434.

14. "Bericht des Zentralkomitees an den VI. Parteitag der sozialistischen Einheitspartei Deutschlands", Neues Deutschland (11. Oktober 1962), 6.

15. See Such, op. cit., note 1.

16. A.H. Rubenstein, "Der Einfluß organisatorischer Faktoren auf Entscheidungsprozesse in Forschung und Entwicklung - Der Fall dezentralisierter Großunternehmen", in J. Naumann (ed.), Forschungsökonomie und Forschungspolitik (Stuttgart: Ernst Klett Verlag, 1970), 349-350.

17. A. Scherzinger, Zur Planung, Organisation und Lenkung von Forschung und Entwicklung in der DDR (Berlin: Duncker und Humblot, 1977), 148 and 166.

18. H. Echterhoff-Severitt, Forschung und Entwicklung in der

Wirtschaft 1971 (Essen: Stifterverband für die deutsche Wissenschaft, 1974).

19. Pöschel, op. cit., note 5, 55.

20. See e.g. L.E. Nolting, The Financing of Research, Development and Innovation in the USSR, by Type of Performer (Washington: U.S. Department of Commerce, 1976), 9-11.

21. Autorenkollektiv, "Die Finanzen...", op. cit., note 1, 176.

22. See Amann, Berry and Davies, op. cit., note 11, 386-390.

23. G. Mittag, "Komplexe sozialistische Rationalisierung - eine Hauptrichtung unserer ökonomischen Politik bis 1970", in Sozialistische Rationalisierung und Standardisierung (Berlin: Dietz Verlag, 1966), 79.

24. N. Moc, "Lange Lieferzeiten bremsen das Tempo der Modernisierung", Die Wirtschaft, Nr. 49 (1966), 5.

25. J. Meyer, "Methodik und Auswertung der Erfassung der Grundmittel und der Arbeitsfläche für Forschung und Entwicklung", Statistische Praxis, Nr. 7 (1967), 411.

26. Ibid.

27. Autorenkollektiv, Das Forschungspotential im Sozialismus (Berlin: Akademie Verlag, 1977), 210.

28. W. Meske, "Entwicklungstendenzen in der Ausstattung der Forschung mit Grundmitteln, Teil II", Die Technik, Nr. 1 (1974), 30-31. For part one of this article see: Die Technik, Nr. 12 (1973), 748-751.

29. E. Honecker, "Bericht des Zentralkomitees der Sozialistischen Einheitspartei Deutschlands an den IX. Parteitag der SED", Neues Deutschland (19. Mai 1976), 10.

30. W. Stoph, Die Durchführung der volkswirtschaftlichen Aufgaben (Berlin: Dietz Verlag, 1967), 29.

31. "Forschungsrat tagte vor dem 5. Plenum des ZK", Die Wirtschaft, Nr. 1 (1964), 4.

32. E. Garbe, "Reserven in knappen Konstruktionskapazitäten", Die Wirtschaft, Nr. 38 (1966), 12.

33. F. Selbmann, "Maßnahmen zur Verbesserung der Arbeit auf dem Gebiet der naturwissenschaftlich-technischen Forschung und Entwicklung und die Einführung der neuen Technik", Die Technik, Nr. 10 (1957), 668.

34. Autorenkollektiv, "Die Finanzen...", op. cit., note 1, 176.

35. "Kurze Überleitungsfristen garantiert Zeit-, Markt- und Produktivitätsgewinn", Die Wirtschaft, Nr. 2 (1966), 5.

36. This imbalance was also noted by the Birmingham group for the first half of the sixties. Amann, Berry and Davies, op. cit., note 11, 387-390.

37. OECD, The Measurement of Scientific and Technical Activities. "Frascati Manual" (Paris: OECD, 1976).

38. Mittag, op. cit., note 23, 79.

39. E. Pässler, "Die industrieverbundene Forschungsarbeit auf dem Gebiet der Fertigungstechnik im Maschinenbau", Die Technik, Nr. 7 (1963), 451 and 453.

40. Such, op. cit., note 1, 127.

41. R. Zander, "Vorschlag zu den Aufgaben eines Industriezweiginstitutes", Die Wirtschaft, Nr. 14 (1966), 6.

42. H. Kerda, "WTZ - seine Funktionen für den wissenschaftlich-technischen Fortschritt und die Rationalisierung", Die Wirtschaft, Nr. 23 (1966), 6-7.

43. H. Kusicka and W. Leupold, "Für eine höhere Effektivität der industriellen Forschung und Entwicklung", Einheit, Nr. 1 (1965), 21.

44. R. Heinze, "Wissenschaftlich-technische Grundkonzeptionen schaffen", Die Wirtschaft, Nr. 39 (1964), 16.

45. Kusicka and Leupold, op. cit., note 43, 22.

46. R. Rompe, "Probleme der Überführung wissenschaftlicher Forschungsergebnisse in die Produktion", Einheit, Nr. 6 (1962), 26.

47. Kusicka and Leupold, op. cit., note 43, 21.

48. E. Rigo, "Kampf gegen Planrückstände und Schaffung des wissenschaftlich-technischen Verlaufs gehören zusammen", Die Wirtschaft, Nr. 23 (1970), 3.

49. Kusicka and Leupold, op. cit., note 43, 22.

50. The FRG sectors "energy/mining" and "manufacturing" were grouped together to approximate the GDR definition of "industry" as given for instance in St.J.DDR 1966, 151. Figures for the FRG's associations were not included as they would not appreciably affect the breakdown. GDR data were adjusted by reducing the category "other employees" by one-third. Occupation and qualifications were compared according to the following scheme:

	Level	GDR	FRG
Qualified Scientists and Engineers (QSEs)	University	Hochschulkader	Wissenschaftler (incl. Dipl. Ing.)
Qualified Technicians (QTs)	Technical College	Fachschulkader	Ingenieure (Ing.grad. and Techniker)

This scheme assumes for the GDR that all university graduates were employed as QSEs, and all graduates of technical colleges, as qualified technicians although it is possible that certain university graduates worked as QTs and certain technical college graduates as QSEs. It also assumes for the FRG that all "Wissenschaftler" had university qualifications, and all "Ingenieure und Techniker", technical college qualifications. West German R&D statistics are collected on a mixed occupation and qualification basis, though the passing of an examination is usually a condition for classification between the personnel groups.
A third point is that in both Germanies there were two main categories of engineering qualification: that awarded by the universities, the "Diplomingenieur", and that conferred by engineering schools, the "Ing.grad.", which was of a lower level. In the GDR, universities conferred the degree "Diplomingenieur" after five years study. "Ingenieurschulen" and "Fachschulen"

awarded lower level qualifications after three years study. See e.g. A bis Z. Ein Taschen- und Nachschlagsbuch über den anderen Teil Deutschlands (Bonn: Deutscher Bundesverlag, 1969), 182-183, 178. The existence of the category "Ing.grad." has been a source of confusion in international comparisons of R&D. In the OECD surveys the West German "Ing.grads." were included under the heading "Technicians" in 1964; under "Scientists and Engineers" in 1967; and in 1969 under both "Scientists and Engineers" (Business Enterprise Sector) and "Technicians" (General Government, Private Non-Profit and Higher Education Sectors). See OECD, International Survey of the Resources Devoted to R&D in 1969 by OECD Member Countries, Statistical Tables and Notes, Vol. 5, Total Tables (Paris: OECD, 1973), 145. In the statistics used for this comparison, the "Ing.grads." are included with "Fachschulkader" for the GDR and with "Techniker" for the FRG. See also Kusicka and Leupold, op. cit., note 2, 45.

51. Autorenkollektiv, "Das Forschungspotential...", op. cit., note 27, 113. The questionnaire appears to have been conducted in 1971.

52. Kusicka and Leupold, op. cit., note 43, 22.

53. Kusicka and Leupold, op. cit., note 2, 53.

54. Ibid.

55. The usual equivalent of the English word "technology" in the GDR is "Technik" and in the FRG "Technologie". A "Technologe" in the GDR was a particular kind of engineer. In enterprises within the machine building sector, he was concerned with the following areas: technological development; technological planning (studies of capacity, enterprise comparisons, measures for improving production technology etc.); technological preparation of manufacture (stipulating the work cycle and the necessary machines etc., material consumption norms, analyses of wastage); work studies, norms, planning and procurement of tools and equipment. See Lexikon der Wirtschaft. Industrie (Berlin: Verlag Die Wirtschaft, 1970), 782-783.

56. E. Bernicke, "Technologie und wissenschaftlich-technische Revolution", Einheit, Nr. 9 (1967), 1112-1113.

57. H. Fritzsche, "Aufgaben der Technologie nach dem 9. Plenum", Die Wirtschaft, Nr. 47 (1968), 3.

58. Bundesministerium für innerdeutsche Beziehungen, Bericht der Bundesregierung und Materialien zur Lage der Nation 1971 (Kassel, 1971), 82.

59. W. Ulbricht, Das neue ökonomische System der Planung und Leitung der Volkswirtschaft in der Praxis (Berlin: Dietz Verlag, 1963), 111.

60. R. Herber and H. Jung, Wissenschaftliche Leitung und Entwicklung der Kader (Berlin: Staatsverlag der Deutschen Demokratischen Republik, 2., unveränderte Auflage, 1964), 119.

61. Ulbricht, op. cit., note 59, 111.

62. Herber and Jung, op. cit., note 60, 122.

63. W. Wolter, in Die Aufgaben der Universitäten und Hochschulen im einheitlichen Bildungssystem der sozialistischen Gesellschaft. IV. Hochschulkonferenz 2. und 3. Februar 1967

in Berlin (Berlin: Staatsverlag der Deutschen Demokratischen Republik, 1967), 130-131.

64. Ulbricht, op. cit., note 59, 111.

65. Herber and Jung, op. cit., note 60, 124.

66. M. Steenbeck, "Macht und Intelligenz im Wandel der Zeit", Einheit, Nr. 6 (1967), 679.

67. Dokumente der Sozialistischen Einheitspartei Deutschlands, Band III (Berlin: Dietz Verlag, 1952), 660.

68. Quoted in A. Schulz, "Erfolge und Aufgaben der Zusammenarbeit mit Produktionsbetrieben", Das Hochschulwesen, Nr. 8-9 (1954), 36-37.

69. Ulbricht, op. cit., note 59, 114.

70. W. Ulbricht, Die Durchführung der ökonomischen Politik im Planjahr 1964 unter besonderer Berücksichtigung der chemischen Industrie (Berlin: Dietz Verlag, 1964), 16-17.

71. St.J.DDR 1976, 62.

72. H. Lysk, "Zielstrebige politisch-ideologische Arbeit beschleunigt wissenschaftlich-technischen Fortschritt", Einheit, Nr. 7 (1975), 716-717.

73. Ibid., 717.

74. All tables labelled with the letter "A" are contained in Appendix A.

75. Based on: OECD, A Study of Resources Devoted to R&D in OECD Member Countries in 1963/64, Vol. 2, Statistical Tables and Notes (Paris: OECD, 1968), 158-159. See figures and definitions in Tables A.7 and A.4.

76. In 1962 12.6 % and 11.6 % respectively. C. Freeman and A. Young, The Research and Development Effort in Western Europe, North America and the Soviet Union (Paris: OECD, 1965), 73.

77. The respective figures in 1960 were 12 % and 14 %, and in 1975, 15 % and 12.5 %. DIW, Handbuch DDR Wirtschaft (Reinbek bei Hamburg: Rowohlt Verlag, 1977), Tabelle 18, p. 313; St.J.DDR 1976, 264. NB: The export figures include building materials.

78. See preceeding section of this chapter.

79. OECD, "A Study of Resources...", op. cit., note 75, 158 and 162.

80. The figures were derived from a printed diagram which may not have been accurately represented, and errors may have been made in measurement. The 1963 R&D statistics were based on a taxonomy used by the National Economic Council. The 1972 figures appear to be based on the system used for industrial production, which in turn was redefined in 1969. An important consequence of these changes was probably the transfer of certain items from the category "machine building" (such as fine mechanics, optics and data processing) to the new category "electrotechnical, electronics and instrument building".

81. W. Ulbricht, Die nationale Mission der DDR und das geistige Schaffen in unserem Staat (Berlin: Dietz Verlag, 1965), 38.

4
The System of Detailed Central Planning, 1945–1962

CENTRAL PLANNING OF THE ECONOMY, RESEARCH AND TECHNOLOGY

After the war the Soviet Military Administration introduced Soviet-type planning methods and institutions into the Eastern Zone of Germany. On 1 October 1945 an order came into effect prescribing that industrial production be managed according to quarterly plans.[1] Longer-term economic plans, disaggregated into operative annual plans, started in 1949 with the two-year plan 1949-1950 and was followd by the first five-year plan 1951-1955, the second five-year plan 1956-(1960) and the seven-year plan 1959-(1965). The most important planning institution was the State Planning Commission, which was officially founded in 1950 although it had effectively existed in various guises from 1947. In the immediate post-war years the plans were concerned principally with replacing damaged and dismantled plants, expanding the iron and steel industry, and generally restructuring industry to overcome the difficulties caused by the partition of Germany. The first five-year plan emphasised the need to raise the technological level of production, and the second stated that by 1960 only "world-level" machines should be manufactured and a high proportion of these should be top-quality products.[2]

The First Party Conference of the Socialist Unity Party (SED) held in January 1949 called for a systematic development of research work within the framework of the two-year plan.[3] The formal planning of scientific-technical work began properly during the first five-year period. The State Planning Commission, assisted by the Central Office for Research and Technology, was responsible for formulating priorities for this work based on the objectives of the economic plan and submitting these to the Council of Ministers. Plans for research and development, the introduction of development results into production, and standardisation were included in the annual economic plans, and an annual "topic list for scientific-technical collaboration with the Soviet Union and the Peoples' Democracies" was set up.

> The development of the peacetime economy of the German Democratic Republic and thus the raising of the population's prosperity requires that technical progress be promoted with every effort throughout the whole of the national economy. In order to achieve this aim, it is necessary to develop the work of the scientists and technicians in the research establishments, laboratories and design bureaux in every respect, and to make the results usable for production without delay.[4]

During the course of the second five-year period the Research Council was created and two new technological programmes were announced. The Research Council was founded in 1957, after the dissolution of the Central Office for Research and Technology, and was attached to the Council of Ministers as advisory committee on scientific-technical research and development. The two programmes were "The Chemical Programme" and "The Programme of Socialist Reconstruction" and aimed at raising labour productivity, officially estimated to be 15 % lower than that of West Germany, and technological levels. The first programme was based on new investments and was part of a programme of cooperation within Comecon. The second stressed better organisation of production and the improvement of existing technology.[5]

The second five-year plan was not completed because the GDR, following the example of the Soviet Union, introduced a seven-year plan in 1959. The "main economic task" of this plan was to raise labour productivity and increase production so as to "catch up and overtake" the FRG in per capita consumption of most consumer goods and food by the end of 1961.[6] Central to the "task" was the acceleration of technological change, and industry had to attain in the shortest feasible period the "world level" of technology.[7]

> If the rapid raising of labour productivity is the key to solving our main economic task, which in turn is closely connected with the struggle for coexistence and the peaceful competition between socialism and imperialism, then science and technology deserve exceptional attention and promotion.[8]

By 1965 labour productivity was to be 85 % higher; the production of the metalworking industry, 118 % higher; and that of the basic materials industry, 90 % higher.[9] Walter Ulbricht stated that every important enterprise in the metalworking industry should set itself the task of beating the leading firm of its sector in West Germany, both with regard to production costs and the technological level of products.[10] In the event, neither the seven-year plan nor the "main economic task" were fulfilled, and according to GDR calculations the productivity gap between the two Germanies actually increased from 15 % in 1958 to 25 % in 1963.[11] Growth rates fell and supply problems appeared. These difficulties, together with the collectivisation measures of 1960, led to increased migration, which was finally stopped on 13 August 1961 by the building of the wall. The plan was dropped de facto in 1961 and annulled in 1963.

In 1961 the State Secretariat for Research and Technology was created to oversee the Research Council and ensure the cooperation of the Council with the State Planning Commission. It was the forerunner of the Ministry for Science and Technology founded in 1967.[12] In 1961-1962 an improved planning methodology for science and technology was also announced and took the form of the "Plan for New Technology". Despite these modifications to the policy-making and planning processes, the party leadership had apparently begun to recognise that the policy of detailed central planning along Soviet lines would not, as hoped, stimulate a rate of technological change far exceeding that under capitalism. The mission of overtaking West Germany had proved to be more difficult than anticipated. The "command economy" had succeeded in achieving rapid industrialisation and militarisation in the USSR, though at considerable economic and social cost. Grafted onto the conditions of the GDR, a country already industrialised, the system was not very effective and the stimulation of technological change was particularly unsatisfactory. The following six sections aim to locate and analyse some of the main obstacles to technological change under the detailed planning system.

OPTIMISM, OPPOSITION AND SCHEMATISM

The planning of technological change in the GDR was characterised by naive optimism, opposition and above all, schematism. Amongst supporters of the regime there was a widespread notion that central planning would efficiently regulate all aspects of the national economy, including technological change. As the physicist and member of the Central Committee, Robert Rompe, stated:

> It has been pointed out with justification time and time again that it is detrimental to believe that innovation in a planned economy takes place without any problems, according, so to speak, to the laws of an inherent logic of planning. There is no automatic mechanism for innovation in the planned economy. Of decisive importance are the people who continually improve the planning and ensure that no essential result of research and development work remains unused.[13]

Many scientists and technologists, on the other hand, felt that the planning of research and development would be a restrictive corset that would inhibit individual initiative, and they pointed out that in the past good results had been obtained in "academic freedom" without a plan.[14] Probably the degree of opposition was related to the prestige of the R&D work and its institution: more opposition is likely to have come from the academic sector than from industry; and within industry, more from the institutes than the enterprise R&D establishments.

Most of those actively involved in the central planning of research and technology lacked planning experience, and many possessed little in the way of relevant technical knowledge. Two consequences of this were poor plan formulation and unsatisfactory supervision of plan execution. In the fifties the planning of research and technology was modelled on production planning: the central authorities enunciated certain aims or targets and enterprises and research institutes were supposed to draft their plans in accordance with these. It was gradually discovered, however, that the amount of planning technique which could be transferred from the management of production to that of R&D was limited. Moreover up to the sixties little attention was paid to the specific features of the various industrial branches. Complaints were made in 1962 that the Plan for New Technology had been based on the conditions and needs of the metalworking industry and that the technical and economic requirements of the basic materials industry, the chemical industry, the light industry and food industry were inadequately provided for.[15]

The party leadership and top-level bureaucrats appear to have expected too great a degree of precision in R&D planning, an extreme but not uncommon view being that all results of scientific research should be "used"; all new designs, introduced into production; and in general all plans for research and technology, fulfilled. Hermann Pöschel, director of the department advising the Central Committee on research and development, and other writers quoted statistics relating to important, so-called "central" or "state" topics in the Plan for New Technology to demonstrate unsatisfactory plan fulfilment for science and technology. Apparently, in 1961 only 122 of the 215 R&D tasks to be completed had actually been finished (56.7 %);[16] only 146 of 219 new products, actually introduced into manufacture (66.7 %);[17] and only 34 of the 57 new mechanisation and automation tasks, begun.[18] In 1962, only 153 of the 206 R&D tasks to be completed had been finished (74.3 %);[19] only 88 of the 112 new products, actually introduced into manufacture (78.6 %);[20] and the individual sections of the whole central Plan for Technology had been only 50-70 % fulfilled.[21] The loss to industry in 1961 from R&D work which was of little or no use to the economy was given as 200 million DM, which was about 15-20 % of all R&D expenditure, and approximately the amount spent on R&D in the chemical and electrotechnology sectors taken together.[22] The number of R&D projects broken off had increased over the years: for every 100 DM of completed R&D work in the electronics sector, there were 9 DM for discontinued tasks in 1958 and more than 21 DM in 1962.[23] There were complaints that the fulfilment of the Plan for New Technology showed no fundamental improvement in the first quarter of 1963, and according to data supplied by the enterprises themselves, almost 21 % of the 500 new products introduced into manufacture lay under the "world level".[24]

Pöschel and other critics were undoubtedly justified in believing that R&D in the GDR was rather slow and that there were considerable delays in getting R&D results into production. What might be questioned is whether tardy completion of the Plan for New Technology was necessarily reprehensible: plan fulfilment surely depends upon the nature of the R&D work and the criterion of "fulfilment". Similarly, one might dispute whether broken-off tasks were always a drain on the national economy: discontinued projects were better than completed but obsolete innovations and such projects may in any case have yielded valuable information. Over the years, naive optimism in planning was probably superseded to a large extent by resigned realism; and opposition, by pragmatic acceptance. Yet as will be seen in the following chapters, schematism in planning persisted into the seventies, especially with regard to improving plan precision.

BUREAUCRACY

During the course of the first five-year plan it became apparent that there was an over-concentration of decision making and an excess of paper work. The higher-level agencies did not have a good overview of the economy and were poorly situated to know what technological changes were necessary or possible in the enterprise. At the 25th Conference of the Central Committee in October 1955, Ulbricht complained that the staff in the main administrations were not informed about the technical deficiencies in production and had not tried to implement governmental decisions concerning technical progress. Money intended to finance the introduction of new technology into the enterprise had been allocated ineffectively; it had been distributed by the ministries among too many enterprises, and spent in part on projects having nothing to do with new technology such as building a new porter's lodge in one Cottbus enterprise.[25]

An attempt was made from 1954 to relieve this situation by permitting the middle and lower levels more economic and administrative freedom. Fewer plan indices were to be prescribed by the centre and the number of state plan positions for so-called "single products", for example, decreased from 950 in 1955 to 810 in 1956 and to 440 in 1957.[26] The centre was instructed to devote its attention to the most important aspects of technological change and not to encumber the central plan with all possible topics. In the chemical industry, the number of central topics for research and development apparently sank from 460 in 1956 to 291 in 1958.[27]

In spite of such figures, the devolution does not appear to have been very successful. Whereas the lower levels sometimes attempted to use their increased rights to oppose the planning and management of science and technology, the upper levels were generally reluctant to relinquish their areas of competence. Ulbricht noted that the combination of centralised management with "economic-operative independence" for the enterprises was in many cases corrupted by employees of the main administrations and ministries "into a bureaucratic centralism, in which works managers are made the errand boys of the Berlin departments, their freedom of movement limited, and their initiative lamed".[28] Moreover, the amount of paper work remained excessive. Ulbricht complained that:

> in the course of research and development there is such absurd bureaucracy which for example requires ... that up to the release of a completed

> development task, 61 administrative actions, consultations, discussions and approvals are to be undertaken.[29]

Pfützenreuter of the enterprise "Carl Zeiß Jena" stated in 1962 that many documents and forms had to be prepared for the superior administrative agencies and this did not exactly facilitate the comprehensibility of the Plan for New Technology.[30] Rompe warned that the amount of administrative work, particularly in the electrotechnical industry, jeopardised creative scientific-technological work and he urged the State Planning Commission, the National Economic Council and the Research Council to take measures to ensure that planning was concerned in the first instance with questions of science and technology, and not with methodological-administrative problems.[31]

PROJECT SELECTION

A common assumption of GDR writings on technological change was that under conditions of socialism, R&D projects could be selected in a more "rational" way than under capitalism. For one thing, the centre could ensure that projects were undertaken which were important for the national economy rather than merely benefitting an individual firm; for another, it could eliminate duplication of effort.

In practice, project selection in the GDR took place less "rationally" than this. Limits on the ability of the centre to exercise effective project choice were set by its administrative distance from the enterprise and institute and by the inflated bureaucracy. In the mid-fifties, none of the central agencies had more than a statistical overview of current development topics. It was virtually impossible to grade proposed R&D projects with respect to their importance to the economy, to decide which should be included in the central plan and which should be left to the responsibility of the enterprise, and to achieve the desired concentration of research topics and elimination of parallel work. Despite the moves to devolve planning from 1954, there were still complaints in 1962 that the plans for new technology were simply compilations of proposals from below.

Contrary to the assumptions of GDR publications, one might argue that the lower levels of the economic hierarchy could best decide about R&D projects since they possessed most of the necessary information, and that parallel projects were not necessarily a waste of money and effort. Concerning the latter point, it has certainly been found in the West that some duplication of effort, in the sense of independent R&D aiming at the same goal, is a useful strategy. The former point is valid, however, only if the lower levels were sufficiently motivated to select and implement R&D projects ultimately beneficial to the economy as a whole. In the previous chapter it was noted that institutes were often rebuked for conducting "hobby research" and in later sections of this chapter, it will be shown that the bonus and price systems did not provide the enterprise with much incentive to transfer into production R&D results involving significant technological advance.

Economic calculations in plan proposals were mainly limited to estimating the R&D costs and applying for the finance needed in the current year: little attention appears to have been paid to the economic use of R&D results. Investigations in one VVB revealed that over a period of five years, approximately 15 million DM of research and development expenditure had been written off without being any use to production. A further investigation showed that in more than 90 % of these cases, the R&D tasks had not been discontinued for technological reasons, but rather for wrong technical-economic estimates.[32] Two reservations about these findings are appropriate. Firstly, the probability of technological completion in the West is also much higher than the probability of market success. Mansfield found in an examination of three U.S. laboratories that for every 100 projects begun, 57 were technologically completed; of those successfully completed, 31 were commercialised; of those commercialised, 12 were market successes.[33] Secondly, a view in the GDR characteristic of the general overestimation of central planning as an implementor of technological change was that better planning would result in better project selection. Criticising the findings for the VVB, the economist Friedrich and his colleague Schulz stated that "to reduce such misdevelopments to a minimum, ... it is imperative to improve the central planning and management of scientific-technical work and of the transfer of its results into production".[34]

In spite of these reservations, GDR literature indicates strongly that the needs of the user and the economy were neglected. In 1959 the party journal Einheit emphasised in an article on electrotechnology, that if the GDR was to catch up with the "world level", more attention would have to be paid to market research. The designer would have to get to know the exact wishes of the customer.[35] In 1961, Professor Fülle, an expert on "material economy", i.e. on supplies and marketing, complained about the lack of market research in industry. Numerous enterprises, he said, particularly in machine building, the fine-mechanical and electrotechnical industries, the chemical and pharmaceutical industries, and the textile and wood industries, were manufacturing unwanted products which could only be marketed with great difficulty.[36] According to Pöschel, whether the results of research and development work were of any economic use was often a matter of chance: R&D workers were content with whatever turned out, and whether a planned task was considered fulfilled or not had depended for a long time on whether the allotted resources had been used up.[37]

Formally, at least, the process of project selection was not restricted to the R&D group and the central planning agency. Other bodies such as defence boards and central work groups were supposed to act as a means of control although they functioned less than satisfactorily. Referring to the work group for basic research, for example, the plasma physicist and deputy director of the Research Council, Professor Max Steenbeck, stated in 1962 that generally money and plan positions for research projects were not insurmountable obstacles, and the examination of a research project often proceeded amongst colleagues according to the motto: "You leave me alone, and I'll leave you alone".[38] Nor was the work of such committees at the enterprise level much better: usually the technical director or R&D manager outlined the tasks of the Plan for New Technology in a global fashion before an enterprise board; often the whole procedure was a mere formality; and since the "defence" of the project was an internal affair, the needs of the future user and the expertise of persons from outside could not be properly taken into account. In effect, projects were frequently selected according to purely internal interests of the institute, the producer enterprise and especially its R&D establishments, to the neglect of the needs of the user and the national economy. Interestingly, this practice was not so different from that found in capitalist firms. Freeman in particular has emphasised that "project estimation" tends to be more "a process of political advocacy and clash of interest groups than sober assessment of measurable probabilities".[39]

PLAN COORDINATION

The synchronisation of plans and plan components was unwieldy. Prior to 1961 there were four types of plan directly concerned with technological change and these often had little connection with the national economic plan. Firstly, there was the Plan "Research and Technology" set for the R&D establishments of enterprises, and the industrial and academic institutes. This was not well coordinated with the plans for production, materials, investment, labour force and finance. Secondly, there was the Plan for Including Finished Designs and Processes in Production set for the departments of the State Planning Commission, the ministries, the VVBs and the enterprises. It contained tasks involving the introduction of R&D projects into manufacture which had been completed the previous year but did not have any close connection to those research and design tasks not yet completed. The remaining plans were the Plan for Standardisation and the Plan for "Technical-Organisational Measures" ("TOM-Plan") and these also lacked close coordination with the other plans for, and related to, technological change. The result was that the various stages from the beginning of research and development work to the commencement of production took place disconnectedly and with insufficient overall supervision. One author characterised the process of technological change in the GDR as a "relay race".[40]

Many R&D topics were begun without the necessary material, personnel and technical capacities, and frequently it was discovered only relatively late that there was no production capacity for the newly developed product. Often the same task for technological change appeared in more than one plan, particularly as there was a lack of coordination between the different agencies responsible for the various plans.[41] This resulted in duplication of effort, part- or non-completion of planned tasks, double notification and wrong estimates of work done.

The planning of materials was too rigid and scarcely appropriate for the uncertainties of research work and the necessity to adapt quickly to changing conditions. According to the law in 1961, all material requirements, including standardised electrical components for machines, had to be ordered by 30 May for the coming year:

> The law therefore clearly obstructs the introduction of new technology. If, for example, a collective of scientists and designers completed a development project in August 1961, then according to the law, a supply of materials for 1962 is no longer possible. Only in 1963 can the order placed for materials in August 1961 be carried out![42]

Paul Verner, candidate for and later member of the Politbureau, criticised the fact that although the "Electrical Apparatus Works Berlin" had developed a series of new switches having a higher technical standard than those of the AEG in West Berlin and had demonstrated them at the Leipzig Fair in Autumn 1959, it was 1961 before the switches were introduced into series production. An important hindrance had been the uncertainty of supply.[43]

Hopes were that by incorporating the main activities of technological change into one plan, the "Plan for New Technology" of 1961, problems of coordination might be relieved.[44] One publication stated sanguinely that this plan would become the main instrument of enterprise management for the fulfilment of plan targets.[45] In 1961 it had eight components as shown in Table 4.1. Particularly important was the perspective plan component for introducing R&D results into production, as it encompassed work from the beginning of R&D up to and including transfer into production. In 1962 the Plan for New Technology was even more elaborate, the idea being that the greater specification of steps involved in technological change would relieve coordination problems and open up the possibility of running certain stages concurrently. Nevertheless there were complaints in 1962 that the plan was poorly coordinated with other plans and that it lacked internal synchronisation. The Plan for New Technology was still not regarded as an important and integral part of enterprise planning, and its contents often did not find corresponding entries in the other enterprise plans. The various component plans overlapped: the raising of quality, for instance, could be included in the Plan for Quality Development and in the Plan "Research and Development", and mechanisation could be included in the Plan for Mechanisation and Automation and in the TOM-Plan. Many enterprises took advantage of this overlap and included a project in more than one component plan, thus more easily fulfilling the Plan for New Technology and appearing to be more active in technological change than was actually the case.

THE ENTERPRISE BONUS SYSTEM

In theory, the introduction of new or improved technology was stimulated by the general enterprise bonus system. A law decree published in March 1962, for instance, stipulated that enterprise bonuses for management and workforce be awarded on the completion of "state tasks" in areas such as production output;

Table 4.1
Component Parts of the Plan for New Technology in 1961 and 1962/Precursor Plans up to 1960

Up to 1960	1961	1962
Plan "Research and Technology"	Plan "Research and Development"	Plan "Research and Development"
Plan for Including Finished Designs and Processes in Production	Plan for Introducing R&D Results into Production - Perspective Tasks	Plan for Introducing New Products into Production - Perspective Tasks
		Plan for Introducing New Processes into Production - Perspective Tasks
	Plan for Including New, Finished Designs and Processes in Production	Plan for Including New Products in Production - Annual Tasks
		Plan for Including New Processes in Production - Annual Tasks
	Plan for Ending the Production of Technologically Old Products	Plan for Ending the Production of Old Products
		Plan for Ending the Production of Old Processes
Plan for Standardisation	Plan for Standardisation	Plan for Standardisation
		Plan for Quality Development
	Plan for Introducing Improvement Proposals and Inventions of Supra-Enterprise Character (only for VVBs and Regional Councils)	Plan for Introducing Improvement Proposals and Inventions of Supra-Enterprise Character (only for VVBs and Regional Councils)
	Plan for Automation	Plan for Mechanisation and Automation
TOM-Plan (= Plan for Technical-Organisational Measures)	TOM-Plan	TOM-Plan

<u>Source:</u> H. Emmerich, "Zur Methodik des Planes 'Neue Technik' 1962", <u>Sozialistische Planwirtschaft</u>, Nr. 7 (1961), 20.

the Plan for New Technology; labour productivity; the lowering of costs; quality; range; and exports.[46] These criteria were to be used by the superior agencies to evaluate the performance, and thus compute the bonuses, of enterprise management. Similarly, enterprise management was supposed to prescribe indices of performance based on these criteria for bonus payments to the various enterprise departments.[47]

In practice, the system of enterprise bonuses discouraged the introduction of new and improved technology. The combined operation of three factors may be viewed as being largely responsible for this. One was that the superior agencies and hence enterprise management regarded production output as the most important task. Another was that enterprise success was not measured in absolute terms but relative to the plan. A third concerned the short length of the operative plan period.

<u>Relative plan fulfilment</u> stimulated an enterprise interest in "slack plans" and a counter interest on the part of the higher administrative agencies in applying the "ratchet principle". Slack plans were ones that could be fulfilled without too much difficulty. It was in the interest of the enterprise to conceal its true capacities from the higher agencies and to try and secure plan targets below its actual capabilities: in this way the enterprise attempted to ensure plan fulfilment and consequently bonus payment. It was more advantageous for an enterprise to have its main target set 5 % higher than the preceding year and to fulfil this 105 % than to plan for an increase of 20 % and only achieve a plan fulfilment of 99.5 %. Although the second case represented a significantly higher enterprise performance, the plan would not be considered fulfilled and there would be no bonuses.[48] Examples from all industrial sectors testified that in the planning of the main enterprise indices, lower figures were used than ought to have been.

> The decisive problem with regard to effecting material interest in the enterprise is to find the right societal standards of measurement for enterprise activity .... Proceeding from the fact that personal and enterprise interests correspond with those of the total society, one would assume that socialist enterprises would be interested in undertaking plan tasks which were as demanding as possible. However, the opposite was often the case. Enterprises were interested in plan targets set as low as possible, as these were easier to fulfil and overfulfilment brought high bonuses.[49]

The "ratchet principle" was the term given to the way the higher level planners tried to "tauten" enterprise plans by using the performance of one plan period as a guideline to setting higher targets in the next. Under these circumstances it was clearly unwise for an enterprise to aim at maximum or even high overfulfilment of the main success index in any particular period. Instead it was more advisable to aim for modest overfulfilment of the plan which left scope for "progress" in future plan periods. This brought regular bonuses, avoided taut plans, and earned the enterprise a good reputation in the eyes of the administration. Although one of the aims of the bargaining process between the enterprise and administrative levels in the setting up of plans was to eliminate "slackness", in practice this was difficult. The higher economic agencies could not know exactly how slack a plan was, and if they set plans which were too taut this would lead to disproportions in supplies.

Taut plans and the ratchet effect obstructed technological change. By definition, taut plans were difficult to fulfil and since the introduction of new technology usually reduced the rate of output, fulfilment of the main plan index was endangered. From the enterprise point of view, significant technological change under these circumstances was best avoided. Only if the project was included in the enterprise plan at the beginning of the plan period, would the enterprise be less averse to introducing new technology. For then the central planners might set lower plan targets in anticipation of lower output, and make provision for supplies. Attitudes of the superior administrations to technological change were probably mixed. On the one hand, they were interested in raising the technological level of their branch and if measures could be undertaken without too much organisational upset in the enterprise, then these would be encouraged. On the other hand, if the introduction of new technology into the enterprise involved problems and in particular loss of output, the administrative agencies would be less enthusiastic for they also had their plans to fulfil; moreover they had to consider the consequences of the loss of output for the customer enterprises and whether alternative supplies could be arranged. Interestingly, although slack plans may be regarded as reflecting enterprise inefficiency, they did furnish more scope for technological change. From this point of view the seven-year plan period from 1959 was not very conducive to technological change since the main macro-economic indicators were set so ambitiously high: the political leadership's hope of achieving a massive increase in output and simultaneously overtaking the FRG in technology was a rather futile one.

The ratchet effect obstructed technological change since it threatened any future benefits to the enterprise resulting from the new technology. Although the enterprise probably had to cope with organisational problems, lower output and lower bonuses on introducing new technology, the planners could take account of this technology for future plans and set higher targets for production, labour productivity, and profitability etc. When, however, the enterprise felt that it either could avoid or had sufficient plan slackness to absorb production delays and a turn of the ratchet, the provision of free state investment funds sometimes stimulated it to acquire new technology as a useful "reserve" for future plan fulfilment and there were reports that in a number of enterprises, new machines were either under-utilised or unused.[50] Apparently only in mid-1962 was the Investment Bank given the formal right to take part in and influence decisions in industry concerning investment plans.[51]

The length of the operative plan period was one year and this was too short to encourage significant technological change. Such technological change often interrupted production and it was difficult to make up the loss before the end of the year. If the plan was taut, so much the worse. Furthermore, since any benefits to the producer from the new technology usually only materialised in later years, these could be excluded by the annual "ratchet" effect.

Although R&D tasks generally stretched over a number of years, projects, at least up to 1961, had to be planned annually, and the associated problems such as procurement of materials, finance, and manpower reconsidered as if the project were starting right from the beginning. As noted above, a perspective R&D plan was introduced in 1961, but given that the other enterprise plans still remained on an annual basis, the new plan is unlikely to have alleviated problems significantly. The effectiveness of other types of longer-term plans for technological change such as "The Chemical Programme", "The Programme of Socialist Reconstruction", and "The Geology Programme"[52] must also be viewed sceptically. Indeed according to Professor Lange of the University for Economics in Berlin, most industrial sectors had not formulated the "basic scientific-technological conception" necessary for setting up perspective plans, and there was widespread unclarity about the contents and possible uses of such "conceptions".[53]

Enterprises tended to "time" their scientific-technological work to the rhythm of the plan-year. They preferred short-term R&D work having low risk, and avoided some of the problems of supply and procurement by getting the projects included in the annual plan. In 1961, more than 80 % of the R&D tasks were planned to be concluded in the fourth quarter of the year. Professor Borchert, director of the Institute for Industrial Economics at the Martin-Luther-University, called in 1962 for greater flexibility concerning the planning period:

> The progress of technology and the winning of scientific findings cannot be forced into the rhythm of the plan year. One cannot decide to do without the use of a technical-organisational measure or a new discovery simply because the relevant section of the Plan for New Technology has just been worked out and completed. For this reason, the Plan for New Technology must be continually revised ...[54]

A longer operative plan period would have facilitated technological change by giving management room to manoeuvre with the introduction of new technology whilst still fulfilling the main plan index. It would also have postponed the discouraging effect of the "ratchet". A new problem would have arisen, however, with regard to bonuses: bonuses paid out every five years or so would have been less of an incentive than those paid out annually.

The main success index for the enterprise was usually "gross production" and this largely determined how high bonuses would be and whether an enterprise would be "honoured" or "distinguished".[55] The index stimulated high output, and enterprises were only likely to be interested in technological change if it did not jeopardise the fulfilment of the production plan. This was a major cause of the delay in transferring R&D results into production, and was probably also a reason why many enterprises did not possess satisfactory technological departments to prepare the results for production.[56] No enterprise, according to a source in 1961, was known to have been "distinguished" for getting a new design into production within a short time period.[57] Nor was it usual for enterprises to be penalised for manufacturing out-of-date products; an enterprise producing technologically obsolete goods but fulfilling or overfulfilling the plan for gross production would, as a rule, receive the customary bonuses and other privileges.

The index for gross production could either be measured as a quantity, a value, or a weight. If the index was in quantity units, the enterprise aimed at manufacturing as many products as possible; quality and range were usually neglected. If the index was given in terms of price, the plan was easier to fulfil when as much expensive material as possible was used. Finally, if gross production was measured in weight, enterprises tried to manufacture in as material-intensive a fashion as possible. An example of this was given by Schenk, a former assistant of the head of planning:

> The Soviet-type planning methodology leads simply to the squandering of national wealth, to indifference and slipshod work .... In machine building, the quantity of production is calculated according to weight. In order to achieve as good a plan fulfilment as possible, many enterprises simply increase the weight of the machines they manufacture. A particularly crass example was reported of a machine tool firm which made the base-plate of a machine one tonne heavier than the design required. It did not worry the firm that as a result steel, which is a scarce commodity in the whole of the Eastern bloc, was being senselessly wasted, that the net capacity of the machine in relation to its weight had deteriorated, and that sales possibilities had become smaller. The enterprise fulfilled in this manner its plan and received its bonuses; everything else was of no interest.[58]

Clearly the success index "gross production" could not help the GDR to achieve high kilogram-prices relative to the FRG for products on the world market.[59] The West German firm had a financial interest in economising on materials. If a GDR enterprise introduced a new or improved product or process which economised on materials then, assuming the number of products it manufactured remained constant, the index gross production would fall when measured in terms of value or weight. If a change of inputs was involved, the enterprise would also face uncertainty about obtaining the necessary supplies within the plan period.

The technological change could, of course, result in savings of labour rather than materials but this also was not in the interests of the enterprise. The savings in labour costs would probably be higher than the increased depreciation costs and on balance the enterprise would have lower production costs. This might have prompted the price agencies to allocate the product being manufactured a lower sales price. If the index gross production was measured in terms of value, then assuming the number of products being manufactured remained constant, the index would appear to drop. In addition, the planning agencies might have been tempted to turn the ratchet and set more demanding plans for the following period. There was a further consequence of a labour-saving technological change: the bonus fund would drop. This was because contributions to the fund were based on the size of planned wages, the maximum fund being 6.5 % of the planned wage fund.[60] Such factors were undoubtedly behind the decision of "EAW Berlin-Treptow" actually to cancel its order for modern production technology because it had managed in the meantime to get hold of more manpower.[61] The reason for the cancellation could not have been lack of finance as this would have been provided from the state budget.

Rewards for technological change offset enterprise interest in production output to only a limited extent. The rewards could be direct, in the form of bonuses paid for scientific-technical work. They could also be indirect, in the form of side conditions attached to full payment into the general enterprise bonus fund. Such side conditions may have been the fulfilment of plans for new technology, mechanisation and automation, or labour productivity etc. Concerning the first type of incentive, one author complained that many enterprises in the fifties simply paid out bonuses to R&D workers on the basis of production plan fulfilment.[62] Other authors criticised that bonuses for scientific-technical work were based too much on age, official position etc. and too little on the actual performance of the collective and the utility of results.[63] The steps reportedly taken in the USSR and Czechoslovakia towards making the size of the reward proportional to the economic usefulness of R&D work had not yet been followed in the GDR.[64] In effect, bonuses for scientific-technical work in the GDR were merely a regular part of the wage.

Concerning the indirect reward for technological change, full payment to the bonus fund could frequently be ensured by arranging for "slack" in the plans set as side conditions. Commenting on the Plan for New Technology, economists Hennig and Reinecke of the Karl-Marx-University in Leipzig stated that enterprises tried to "plan" only those measures they were sure could be realised. In fact the term "measures" was something of a misnomer, for often the enterprise proposed only general aims (such as reducing costs) without specifying concrete solutions.[65] The Plan for New Technology could then be completed by drawing on "hidden reserves". Other ways of ensuring "slack" were to include the same project in more than one plan component, and to propose relatively riskless technological changes introduced and tested by other enterprises. If an enterprise decided or was directed to make a significant technological change, fear of the ratchet could prompt it to bring out a "non-optimal" design at the beginning, so as to allow scope for "improvements" later. Similarly with "small" technological changes, the enterprise was wise to keep a portfolio of improvements and bring out a few each year to indicate "progress": too many at once might have resulted in a price drop.

Frequently, however, non-fulfilment of the side conditions does not seem to have entailed substantial reduction of the enterprise bonus fund. In machine building, for instance, non-fulfilment of the Plan for Including New Products in Production could incur a reduction of the enterprise bonus fund of only 0.25 %.[66] As a consequence, many enterprises were not overly concerned about fulfilling such plans. The "Berlin Incandescent Lamp Factory" made "extraordinary efforts" in the second half of 1961 to catch up on a large production deficit and it succeeded in fulfilling the output index. The plan component Technical-Organisational Measures ("TOM") was apparently only 60 % fulfilled with regard to the number of measures, and 40 % with regard to the "usefulness". Even after the activity of the whole enterprise party organisation had been oriented towards the fulfilment of the Plan for New Technology, the most important projects of the TOM-Plan for the first quarter of 1962 were again carried through late or not at all.[67] Similarly, in the "Electrochemical Combine Bitterfeld", the final date for drawing up the plan component TOM for 1962 was not kept by various departments. The factory "North" should have submitted the TOM-Plan to the planning department of the combine by 30 August 1961 but it still had not done so by mid-November. The accounting of the TOM-Plan was often delayed in the factory and so no proper check could be made on the fulfilment of tasks.[68]

Paradoxically, perhaps, plan indices such as "mechanisation level" and "labour productivity" could of their own account strongly discourage the introduction of new technology. The index "mechanisation level" was defined as the ratio of the number of production workers employed on machines to the total number of production workers, and was usually expressed as a percentage. Hennig and Reinecke pointed out that such global indicators could be no orientation to the enterprise, and gave the following illustration. If the number of production workers employed on machines equals 600, and the total number of production workers equals 800, then the mechanisation level is 75 %. Assuming that measures for mechanisation and part-automation are undertaken which result in the saving of 120 production workers formerly employed on machines, then the mechanisation level comes to (480/680)x100 which is 70.6 %. In this case, the calculated mechanisation level appears to have sunk, even though the level has effectively risen.[69]

The most commonly used index for labour productivity was the quotient of gross production at fixed plan prices and expenditure of labour. Economist Siegfried Heyde, amongst others, observed that this indicator led to spurious results, and could hinder technological change. It dropped in the textile industry on the changeover from wool to spun rayon. It also dropped in the foundry industry if customers wanted supplies of thin cast iron, as more labour was required to produce one tonne.[70] Heyde noted some reasons why one enterprise had exhibited falling labour productivity over the previous three years. One was that the enterprise had taken over a foundry having 200 employees: since the foundry was producing mainly for the internal needs of the enterprise, the enterprise's output remained roughly constant but its manpower had increased by 200. Another reason was that 60 toolsetters who had previously been classified for planning purposes as "technical personnel" were regrouped as "assistant production workers". A third reason was that a larger volume of spare parts were manufactured in 1961 than in earlier years, and wage costs for this work were higher than for normal production.[71]

Leading figures frequently criticised the "Cinderella" status of plans for technological change. Ziller, the secretary of the Central Committee, said in April 1955 that the main obstacle to technological change was that the ministries and main administrations did not regard the planning and continuous promotion of science and technology as one of their most urgent tasks.[72] A similar charge was made in July of that year by the Council of Ministers in the preamble of a law concerning scientific-technical progress.[73] The deputy chairman of the State Planning Commission, H. Grosse, complained in 1960 that enterprise management was insufficiently interested in questions of scientific-technological development, and that such questions were often left to a handful of specialists.[74] Rompe censured the practice of using R&D personnel to catch up on the production plan,[75] and Steenbeck stated that scientists who were genuinely interested in the practical use of their results often resigned.[76]

PRICES

The price system reinforced the bonus system in discouraging the enterprise to pursue technological change. As Helmut Mann, director of the Institute for Prices at the University for Economics in Berlin noted:

> From the operational point of view, a factory manufacturing technologically obsolete products works in many cases more profitably than it does when it introduces new products into production. Moreover, the manufacture of old products results in a better fulfilment of the other plan indices .... From the point of view of the national economy, however, the situation is the reverse.[77]

Profitability, it is true, did not have the same status among enterprise indices as "gross production". Conceivably, an enterprise could make a profit and yet receive little in the way of bonuses since it had failed to fulfil the production plan. Nonetheless, it was important for the enterprise to make a profit so as to have funds to fulfil the production plan and pay out bonuses.

One reason for the fall in enterprise profitability on introducing a new product, and hence enterprise resistance to technological change, was the often high start-up costs. These were highest in the first year of production. Part of them were paid by the state through the investment plan. The remaining costs had to be covered by the enterprise through a charge in the price calculation of the new product, and since these could be recouped only after a number of years, they had to be provided in the first instance from working capital or bank credits.[78] Another source of finance from 1959 was the "Fund for New Technology" created at enterprise level and containing proceeds from the sale of trial production, prototypes, testing models and pilot lots. Two limitations on the use of this fund for financing start-up costs were, however, that it could not exceed two percent of the enterprise wage bill, and was also used to finance bonuses for technological change.[79]

Even more important reasons for the fall in enterprise profitability on implementing technological change resulted from the fact that prices were fixed and, as a rule, based on prime costs. Fixed prices meant that as production costs declined over time through "learning by doing" and economies of scale, the profit margin increased. As a rule, therefore, older products were more profitable to manufacture than newer ones. An illustration of this is given in Table 4.2 which shows the price components for a

Table 4.2
Development of Price Components for Vacuum Cleaner Motor

	1955	1956	1957	1958	1959	1960	1961
Material	12	12	11.69	11.18	10.18	9.63	9.16
Wages	6	5.50	5	4.80	4.67	4.07	3.80
Overhead Costs	20	18.60	17.60	16.60	16.10	15.70	14.97
Profit	2	3.90	5.71	7.42	9.05	10.60	12.07
Factory Price	40	40	40	40	40	40	40

Source: "Falsche Festpreise hemmen die Produktivität", *Die Wirtschaft*, Nr. 5 (1961), 5.

vacuum cleaner motor over the period 1955-1961: profit on the motor rose by a factor six within six years. Fixed prices gave no incentive to introduce new products into manufacture, nor to make product or process improvements if these involved increased costs.

The attractiveness of manufacturing old products under the fixed price system may have been weakened by two sorts of administrative intervention: general and specific price revisions. General price revisions did not take place frequently owing to the complexity of the task. Immediately after the war, prices were frozen at the level of 1944. With the introduction of longer-term central planning from 1949, certain exceptions to this rule were made and prices permitted on the basis of individual enterprise costs. In 1953, all prices (except for basic materials) were recalculated on the basis of branch-average costs. This system remained in force up to 1964. Theoretically it was possible to revise the price of a specific product making too much profit owing to obsolescence. In practice this is unlikely to have occurred as often as was necessary. On the one hand, enterprises were reluctant to declare their full profit margins and on the other hand, the price agencies were overworked and especially in the fifties suffered from a lack of experienced personnel with the requisite technical knowledge. Price revisions might have affected price relationships negatively in other sectors of the economy. Frequent price changes would have led to a loss of information for the higher economic levels, making detailed planning difficult if not impossible. Stability of prices, moreover, had a considerable attraction over such characteristics of the capitalist system as inflation, recession and unemployment.

Cost prices were based on enterprise prime costs and incorporated a fixed profit mark-up of 6 %. For established products, such prices were calculated from the average prime costs of a group of enterprises. The prices of most new products, however, were based on individual enterprise costs. As in the Soviet Union, many GDR theoreticians considered cost prices to be a correct application of Marx's theory of value, i.e. the theory that prices should reflect the socially necessary expenditure of labour. At the same time these prices constituted a considerable obstacle to technological change since producers were rewarded neither for new products of lower cost nor for increased utility to the user, i.e. cost prices did not reflect use-value.

Consider the case when two similar products could be produced at approximately the same cost. Cost prices alone gave no direction to the producer enterprise to choose the variant having the best technical and productivity characteristics. As Schauer noted:

> Cost prices conserve existing technology. Whether the capacity of a new product has increased by 10 % or 50 % with the same or only marginally increasing costs; whether the operating costs for the user decrease by 5 % or by 30 %, these changes are not expressed through cost prices.[80]

Consider also the case of a new material-saving development which cost less to produce than a similar product. According to the rules for price formation, the customer should receive a lower price, yet this reduced the manufacturer's profit margin in two respects. One was that the profit for the new product was less than the initial profit for the old product since the new prime costs were lower.[81] Another was that the profit margin for the old product had increased anyway owing to the effect of fixed prices and declining costs. The manufacturer suffered also from lower values for the indices "gross production" and "labour productivity" because these were often measured in price and because production changeovers caused a fall in output. The result was lower bonuses. Only when the price of a new product had been fixed by the price agency did the enterprise have an interest in low costs. Of course the enterprise had little interest in its cost-reduction measures leading to the product being classified as "new" for this could have involved a price reduction.

The nature of cost prices in fact stimulated enterprises to "overestimate" the prime costs of a new product in an attempt to obtain the same profitability as that of the old product.[82] This was undoubtedly the reason why the prices for some machines were higher than those of the original models, in spite of lower expenditure of material and labour.[83] One means of combatting arbitrary new cost prices was for the price agencies to compare the price submission with the existing prices of enterprises manufacturing similar products.[84] In practice this was only possible for a limited number of products and so although the practice of "overestimating" costs was frowned upon by the central authorities, it did perhaps facilitate technological change more than "honest" cost estimates. The more "honest" an enterprise was in stating costs, the more difficult its later auxiliary plan fulfilment would be (e.g. "technical improvements", profitability etc.). Even if all enterprises were honest, the resulting price would still depend on the efficiency or inefficiency of the particular enterprise.

The problem of cost prices was acknowledged by GDR writers and different methods of price formation were discussed, though not used to a great extent. These methods were known in the literature as "relation prices", and the general idea was to price a new product in relation to others. Relation prices had two main advantages. Firstly, they were not so arbitrary as the individual cost prices. They guaranteed that the price of a new product would be consistent with the prices of similar products and could counteract the practice of "overestimating" individual costs. The method was most effective when similar products of a product group were involved rather than the products of one enterprise. Nevertheless, although consistency was guaranteed, the method could not avoid projecting faulty price levels onto the new product. Secondly, an enterprise which managed to produce a new product at lower cost than comparable products was not automatically penalised by receiving a lower price and thus a lower profit, lower values for "gross production", "labour productivity" etc. as with cost prices. Whether this was actually the case depended on which type of relation price formation was used. Relation prices may be divided into two classes, "parametric" and "calculation", and it is not difficult to find examples from each of these classes which did penalise the producer for lower costs. The "weight price", for instance, was a crude type of parametric price used in heavy machine building and meant that a new lighter machine had a lower price than the old machine.[85] The "building block system" was a "calculation price". The final price was often the sum of standardised component prices and if a producer changed the design of a product and saved on components, the total price would drop. So too would his profit.[86]

If the practice of pricing in machine building was typical of the whole of industry, then it would appear that relation prices did not reward the producer any more than cost prices for increased benefit to the user. Conceivably, a parametric price could have been contrived to include user parameters but as a rule, producer parameters were usually used and these were mainly cost parameters.[87] An important reason for this was of course the concern that prices should conform with Marx's theory of value by reflecting the societal expenditure of labour. Function parameters were of particular interest to the user but were not closely related to the societal expenditure of labour. As the economist Helmut Mann noted, societal labour expenditure could not be expressed in Watts, Amperes and other technical quantities.[88] With parametric prices, enterprises had an incentive to concentrate on those parameters having a high coefficient in the price calculation and this was not necessarily advantageous to product quality and costs. Possible product improvements which would negatively affect the parameters for the enterprise were unlikely to be realised. A further problem was that start-up costs were not provided for with these prices, since the prices were based in relation to others.

In the early sixties, roughly 80 % of the new products in the machine building industry were allotted prices based on individual enterprise costs. 15 % of the new products were given calculation prices; and the remaining 5 %, parametric prices.[89] There were perhaps two reasons why cost prices were so predominant. One was doctrinal; the other was that they were easier to calculate than most other methods. This desire for simplicity was also apparent with parametric prices: even when several parameters ought to have been involved in the price construction, in practice only one or two were selected.

A pragmatic attempt to alleviate the difficulties posed to technological change by the price mechanism should be mentioned. This involved the adjustment of enterprise profit: in 1959 and 1960 laws were passed which aimed at compensating the enterprise for the fall in profit owing to the introduction of new technology. The compensation took the form of a supplement added to total enterprise profit (as opposed to being included in the price) and corresponded to the most favourable profit margin of similar products or, when comparable products were not available, lay between the average profit of all the products of the enterprise and the profit of that product with the largest profit margin.[90] The laws did not, however, provide a penalty for enterprises not engaging in technological change.[91] At the end of 1961, therefore, a further law was passed to permit the reduction of enterprise profit for those enterprises which either continued to manufacture products listed in the Plan for Ending the Production of Old Products, or manufactured products not meeting stipulated quality levels, or failed to reach the targets of the Plan for Including New Products in Production.[92]

In practice the method of adjusting enterprise profit did not succeed in overcoming the problems posed to technological change from the price mechanism. An important drawback was that the producer still received no reward for increased utility to the user. The enterprise profit supplement had several other shortcomings. Firstly, it was administered by the VVB, and enterprises were probably not very enthusiastic about declaring their real profit margins for the purpose of calculating the supplement, particularly if the supplement was intended for another enterprise. Secondly, the period of operation of the enterprise profit supplement was one year, which was too short: even if an enterprise was adequately compensated for the fall in profit in the first year, no compensation was available for subsequent years. Thirdly, no compensation was allowed for the deterioration of indices such as "gross production", and "labour productivity". Had the supplements been included in the price rather than enterprise profit, this last difficulty could have been surmounted.

There were also weaknesses of the enterprise profit deductions. One was that the VVB personnel did not possess detailed knowledge of all their enterprises, and hence could not always know when deductions were applicable. It is also doubtful whether the administrators were as ready to impose profit deductions on their enterprises as they were to grant profit supplements: they would have had little interest in being responsible for a large number of apparently unprofitable enterprises. Moreover, there were a number of potential loopholes in the laws so that the imaginative enterprise could find an excuse for avoiding the deductions. It was stipulated for instance that enterprise profit deductions should not jeopardise the measures of the GDR to protect its economy against the "disruptive efforts of West German military circles". And profit cuts could be made neither for spare parts needed for existing fixed assets, nor for export goods.[93]

The foregoing discussion on prices has concentrated mainly on the consequences for technological change resulting from the methods of price formation, and has only touched upon another aspect of prices which also hindered technological change: price administration. As with the planning of technological change in general, there was too much upper level management and too much paperwork. Over the period 1945-1962, a number of changes were made in price administration, particularly at the upper levels. After the founding of the GDR, the top price agency was the Ministry of Finance and this supervised the various regional price agencies. In 1950, central departments were formed for the most important branches of production and these took over much of the work in price formation. Apparently this resulted in too little an influence on prices being exerted by the other agencies responsible for industry. In 1953, when the government promulgated a law on introducing a unified system of fixed prices, it also reorganised the pricing agencies. Responsibility for working out prices in the centrally managed part of the economy was transferred from the central departments to the industrial ministries. Responsibility for pricing in the regional economy was transferred to the regional councils, and that for the private economy, to the Ministry of Finance. The State Planning Commission was supposed to ensure that price measures did not conflict with the "main economic proportions", and the Ministry of Finance was to deal with "finance-political questions". Nevertheless, there were still difficulties: the ministries lacked coordination; there was an overlap of competence between the State Planning Commission and the Ministry of Finance; and there was insufficient cooperation between the various agencies responsible for the state and private sectors, and the centrally and regionally managed enterprises.

In 1956, therefore, a new decree was published ordering a Governmental Commission for Prices to be formed, and pricing for all enterprises of a branch, irrespective of their ownership and whether they were managed centrally or regionally, to be transferred to the industrial ministry responsible for that branch. In 1958 changes were made again, the most important being the abolition of the industrial ministries in imitation of Khrushchev's <u>sovnarkhoz</u> reform. Their responsibilities were taken over by the State Planning Commission which again were transferred in 1961 to the National Economic Council.

It would be beyond the scope of this section to examine in detail the effect of all these changes on pricing activity, but it would be plausible to assume that whereas a good deal of effort went into reorganisation, this was largely reorganisation of the wrong kind. It is doubtful whether there was much reduction in the amount of form-filling, whether there was much acceleration in the price confirmation procedure, and whether there was a better coordination and delineation of the various competences. One kind of reorganisation which might have resulted in such improvements was rejected firmly:

> Some economists in the German Democratic Republic held revisionist views aimed at abolishing the principle of democratic centralism in price formation. The idea behind these damaging views was to introduce a "self-administering economy" and hence permit a free price formation according to so-called "market economy" requirements and dismantle state management. This led to a complete revision of the Marxist-Leninist theory of the state. The realisation of this idea would have been extremely dangerous for the construction of socialism in the German Democratic Republic.[94]

Taken together, the price and bonus systems; schematism and opposition in planning; bureaucracy; problems of project selection; and difficulties in plan coordination amounted to a formidable pressure against technological change. Whilst this list of factors does not pretend to be exhaustive, it does go some way towards explaining why the GDR did not keep pace with, let alone overtake, the FRG in technology. If there had been no countervailing pressures in favour of technological change, the GDR would probably have fallen even further behind the FRG than it actually did.

PRESSURES FOR TECHNOLOGICAL CHANGE

Perhaps the most fundamental pressure for technological change was the economic competition between the two Germanies. Inhabitants of the GDR were fairly well informed about the standard of living in the FRG, particularly through family ties and the reception of radio and television programmes. Engineers were aware of the level of West German technology and could acquaint themselves with the latest developments through FRG publications, the import of FRG products, and the exhibition of FRG goods at the Leipzig Fair. This competition was of course explicitly recognised by the party leadership, and they together with the central authorities exerted pressure in at least three related ways: through directives and plans; through "ideological" emphasis; and through the various mass movements.

Directives and special plan components may have resulted from top-level committees and conferences on problems of industry, or have been based on the expertise of persons in the planning agencies, the Research Council, the suggestions of R&D institutes, and the needs of user enterprises. Influence could also have been exerted by administrative agencies concerned with product quality such as the German Bureau for Measurements and Quality Control.

Directives could force the adoption of new products and processes, compel industry to employ scientists and engineers, and perhaps most importantly, lead to the creation or redeployment of enterprises, institutes, and VVBs. Examples of the latter form of directive may be seen in the development of the electronic data processing industry in the GDR. Most R&D in this field had initially been conducted by the Technical University of Dresden and "Carl Zeiß Jena", but from 1957 a number of central directives were issued aimed at expanding capacity in research, development and production. In 1957 the "VEB Electronic Calculating Machines" was created and in 1960 two institutes were set up: the "Institute for Administrative Organisation", Leipzig, and the "Central Institute for Automation", Dresden. In 1964, the television factory "VEB Rafena Radeberg" was instructed to change over to computer production, the "VVB Office Machines" was transformed into the "VVB Data Processing and Office Machines", and this VVB was also given charge of the Central Institute for Automation (now renamed as the "Institute for Data Processing").[95]

Although industrial management did not usually rank plans for technological change very highly in their scale of priorities, sometimes the leadership and central agencies may have attached especial importance to the fulfilment of the Plan for New Technology and have been relatively unconcerned in the first instance about high production output. In such cases they might have tempted a greater spirit of enterprise innovation by basing the main enterprise bonuses upon the introduction of new technology, by allocating slack output plans, and by guaranteeing supplies. An example along these lines was cited by the economist Werner Kalweit, who noted in 1962 that the "Industrial Works Karl-Marx-Stadt" allocated 60-70 % of its bonus fund for the reward of technological change: the remainder of the fund was used to reward current production.[96]

"Ideological" emphasis of technological change took several forms. Perhaps the most obvious was the pronouncement of leading political figures on the need to "catch up" West Germany in technology and labour productivity and the attention devoted by the media to this theme. Another form of ideological pressure was through the bestowal of special awards to individuals, "collectives", "part-collectives" and enterprises. These honours were various but fell into four main categories: orders, such as the "Karl-Marx-Order" and the "Banner of Labour"; prizes such as the "National Prize" and the "Design Prize"; medals such as the "Humboldt Medal" and those for "Outstanding Performance" in various industrial sectors; and honorary titles such as "Hero of Labour", "Exemplary Inventor" and "Outstanding Scientist of the People". These awards were of a "moral" character, but sometimes also included a financial reward.

A rather less agreeable type of ideological pressure was the public censure of enterprises or VVBs, especially by the party leadership, at important high-level conferences and congresses. When Ulbricht first announced the "main economic task" at the Fifth Party Congress in 1958, for example, he expressed dissatisfaction about the level of technology in GDR industry, and publicly censured some enterprises and branches by name.[97] He complained that for years in the branch "automatic control technology" many designs had got no further than the prototype stage owing to insufficient willingness on the part of enterprises to introduce new products; that in the "VVB Electrical Apparatus" at least twenty completed developments had not been transferred into production; and that the "High Grade Steel Works '8 May 1945', Freital" manufactured poor quality steel and failed to keep delivery dates. Investigations in the "Camera Works Dresden-Niedersedlitz" and "Cinema Works Dresden" had revealed, he said, that the technological lag behind the "world level" had actually increased. The GDR had not managed to produce semi-automatic cameras with diaphragm coupling, even though they had been available for some time in the capitalist countries. Slide projectors with semi- or fully automatic controls were still not manufactured. The GDR projector type "AK 16" was nearly 40 % heavier than the Swiss-made "Bolex H16". The new 16 mm projector "LMP" from Zeiß required five cases for transport and weighed about forty kilos whereas the American "Kodascope Pageant" fitted into one case and weighed only eighteen kilos. Furthermore, the Zeiß projector had a complicated film passage and made a disturbingly loud noise. Ulbricht reproved the director of the "Camera Works Dresden" for lack of interest in pursuing new developments, and the staff of the State Planning Commission for being "led by the nose" by such managers.

The third way in which the centre could exert pressure for technological change was through mass movements such as the party organisation, the trade unions, "the activist movement", the "innovator movement" and the Chamber of Technology. The party organisation was intended to help implement decisions of the Central Committee at all levels of the economy, and at enterprise level was supposed to devote particular attention not only to fulfilment of the production plan but also to raising labour productivity, to rationalisation and to the transfer of science into production.[98] It also gave the cue to the other mass movements, whose duties to a certain extent overlapped: the activists were concerned with raising labour output and norms, the innovators with encouraging rationalisation proposals from the workforce, and the Chamber of Technology with orientating the work of GDR engineers in the direction of the economic plans.[99] These mass organisations may have counteracted the reticence of enterprise management to implement technological change to some degree. That they were an effective pressure must be doubted when one considers the inherent conflict of interests. The enterprise party management, for example, was charged with overseeing technological change, but at the same time had to help ensure production plan fulfilment. The trade unions were in a similar situation with the additional problem of being supposed to represent the interests of the workers. Within the other organisations there was a problem of, on the one hand, permitting the individual to exercise creative initiative, and on the other, organising people collectively to fulfil a plan. Thc innovator movement, for instance, was supposed to tap the creative potential of the workers, yet in the early sixties it was rebuked for being too spontaneous.[100] In later years the movement increasingly became an organisation which completed tasks planned from above.

One pressure for technological change different from the institutionalised ones emanating from the centre was individual initiative. Many individuals, even if rejecting much of what the Socialist Unity Party stood for, probably identified with the leadership's wish to catch up the FRG in technology. Individual initiative could have played a significant role in by-passing the inflexibility and lack of incentive of the planning system. R&D workers might have championed their topics right through to manufacture. Enterprises could have complemented the plan by a system of informal contacts, particularly with regard to securing "extra-plan" supplies from elsewhere in order to overcome bottlenecks and shortages. Enterprise managers may sometimes have been motivated by wider, more "socialist" considerations than simply production plan fulfilment, or might have felt that a number of successfully implemented technological changes would be a useful asset in career prospects. A personal quality which was useful, if not essential, for managing technological change in the GDR was artfulness. The enterprise manager who studiously avoided taut plans and the ratchet effect was in a much better position to implement technological change than his colleague who honestly informed the planning authorities of his capabilities. Similarly, the manager who inflated his cost estimates was more likely to bring out a new product than one who did not manipulate costs.

The sources and channels of pressure towards technological change in the GDR were various, but the body of evidence persuades one to conclude that these were outbalanced by the obstacles. The GDR had failed in its mission to overtake West Germany in labour productivity, technology and living standards. This does not imply that technological change in the FRG involved no problems of planning, bureaucracy, project selection, prices and rewards etc., nor does it suggest that the majority of West German firms pursued policies of radical innovation: owing to the high risks inherent to radical innovation in capitalism, much of the research and technological activity of firms is devoted to relatively minor changes and modifications.[101] Nevertheless, it is the case that competition in the FRG has served as a crucial impetus against technological inertia. Whereas the political leadership had previously chosen to blame "wrong" ideological attitudes, to condemn or persecute potential reformers (such as Benary and Behrens) and to stress the necessity of perfecting the planning mechanism, in the early sixties it began to look for an alternative policy which would stimulate greater industrial efficiency and a higher rate of technological change. As Professor Kalweit noted, an isolated solution of this or that question was inappropriate and a "patchwork" of single economic measures would help no further. What was necessary was to review the entire system of planning and management, and to change it.[102]

NOTES

1. H.-H. Schober, "Die zweite Komponente der Planwirtschaft", Einheit, 2, Nr. 6 (1947), 540.

2. Direktive für den zweiten Fünfjahrplan zur Entwicklung

der Volkswirtschaft der Deutschen Demokratischen Republik 1956 bis 1960 (Berlin: Dietz Verlag, 1957), 7, 34-49. On GDR terminology see e.g.: R. Hanicke, "Wissenschaftlich-technischer Fortschritt, Weltniveau und Höchststand", Einheit, Nr. 7 (1964), 128; and "Weltniveau", in Lexikon der Wirtschaft. Industrie (Berlin: Verlag Die Wirtschaft, 1970), 855.

3. Dokumente der Sozialistischen Einheitspartei Deutschlands, Band II (Berlin: Dietz Verlag, 1950), 190 ff.

4. From the law for the first five-year plan quoted in Forschung und Technik, Schriftenreihe Der Fünfjahrplan, Heft 2 (Berlin: Verlag Die Wirtschaft, 1952), 12.

5. "Entschließung des 5. Plenums des Zentralkomitees der SED", Neues Deutschland, Ausgabe B (26. Mai 1959), 3.

6. "Gesetz über den Siebenjahrplan zur Entwicklung der Volkswirtschaft der Deutschen Demokratischen Republik in den Jahren 1959 bis 1965", in W. Ulbricht, Der Siebenjahrplan des Friedens, des Wohlstandes und des Glücks (Berlin: Dietz Verlag, 1959), 159.

7. Ibid., 166-167.

8. Protokoll der Verhandlungen des V. Parteitages der Sozialistischen Einheitspartei Deutschlands (Berlin: Dietz Verlag, 1959), 86.

9. "Gesetz über den Siebenjahrplan", op. cit., note 6, 166 and 171.

10. Durch sozialistische Gemeinschaftsarbeit zum wissenschaftlich-technischen Höchststand im Maschinenbau und in der Metallurgie, Materialien der 9. Tagung des ZK der SED (Berlin: Dietz Verlag, 1960), 7.

11. Quoted in G. Leptin and M. Melzer, Economic Reforms in East German Industry (London: Oxford University Press, 1978), 6.

12. For formal details on the central institutions concerned with policies and plans for R&D and on the R&D planning process see: A. Scherzinger, Zur Planung, Organisation und Lenkung von Forschung und Entwicklung in der DDR (Berlin: Duncker und Humblot, 1977).

13. R. Rompe, "Über die Zusammenarbeit zwischen Wissenschaft und Produktion", Einheit, Nr. 6 (1959), 723.

14. See e.g. P.A. Thießen, "Probleme der Planung und Organisation der wissenschaftlich-technischen Forschung und Entwicklung in der Deutschen Demokratischen Republik", Einheit, Nr. 5 (1959), 595; M. Steenbeck, "Macht und Intelligenz im Wandel der Zeit", Einheit, Nr. 6 (1967), 682.

15. A. Lange, "Den Plan 'Neue Technik' zum Hauptinstrument des wissenschaftlich-technischen Fortschritts machen", Wissenschaftliche Zeitschrift der Hochschule für Ökonomie, Berlin, 7, Nr. 1 (1962), 5.

16. H. Pöschel, Leitung von Forschung und Technik und wissenschaftlicher Meinungsstreit (Berlin: Dietz Verlag, 1964), 53. In a different publication, Pöschel states that of the 57 R&D tasks to be completed for the central plan, only 30 were finished on time, and of these only 12 were made ready for production. The apparent contradiction between the two sets of data perhaps

results from Pöschel referring in the second case to a particular section or quarter of the central plan. C.f. H. Pöschel and S. Wikarski, "Durch ein System der einheitlichen und straffen Leitung zu einem hohen Nutzeffekt der wissenschaftlich-technischen Arbeit", Einheit, Nr. 2 (1962), 35.

17. Pöschel, op. cit., 54.

18. H. Emmerich, "Für die zentrale Planung des technischen Fortschritts", Sozialistische Planwirtschaft, 3, Nr. 5 (1962), 4.

19. Pöschel, op. cit., note 16, 53.

20. Pöschel, op. cit., note 16, 54.

21. H. Pöschel and S. Wikarski, "Ideologische Probleme bei der Lösung der vom VI. Parteitag gestellten Aufgaben in Wissenschaft und Technik", Einheit, Nr. 6 (1963), 63.

22. Pöschel, op. cit., note 16, 29.

23. Pöschel and Wikarski, op. cit., note 21, 69.

24. Ibid., 63.

25. W. Ulbricht, Die Rolle der Deutschen Demokratischen Republik im Kampf um ein friedliches und glückliches Leben des deutschen Volkes (Berlin: Dietz Verlag, 1955), 42.

26. B. Leuschner, "Aktuelle Probleme unserer Volkswirtschaft", Die Wirtschaft, Nr. 51 (1956), 6.

27. J. Roesler, "Planung und Leitung des wissenschaftlich-technischen Fortschritts der DDR in der zweiten Hälfte der fünfziger Jahre", Jahrbuch für Wirtschaftsgeschichte, T III (1970), 53-54.

28. W. Ulbricht, Fragen der politischen Ökonomie der Deutschen Demokratischen Republik (Berlin: Dietz Verlag, 1955), 17.

29. W. Ulbricht, "Über die Arbeit der Sozialistischen Einheitspartei Deutschlands nach dem XX. Parteitag der KPdSU und die bisherige Durchführung der Beschlüsse der III. Parteikonferenz", in Zur sozialistischen Entwicklung der Volkswirtschaft seit 1945 (Berlin: Dietz Verlag, 1960), 581.

30. W. Pfützenreuter, "Die Verbesserung der Arbeit mit dem Plan Neue Technik", Deutsche Finanzwirtschaft, Ausgabe Finanzen und Buchführung, Nr. 12 (1962), F12.

31. R. Rompe, "Probleme der Überführung wissenschaftlicher Forschungsergebnisse in die Produktion", Einheit, Nr. 6 (1962), 30.

32. G. Friedrich and W. Schulz, "Zu einigen Aufgaben der VVB bei der Leitung des technischen Fortschritts in der Industrie", Einheit, Nr. 6 (1962), 33.

33. E. Mansfield, Research and Innovation in the Modern Corporation (London: Macmillan, 1971), 56-57.

34. Friedrich and Schulz, op. cit., note 32, 34.

35. "Bericht über eine Beratung zu Problemen der Elektrotechnik in der Deutschen Demokratischen Republik", Einheit, Nr. 11 (1959), 1505.

36. H. Fülle, Der Absatz der Erzeugnisse unserer sozialistischen Industrie (Berlin: Volk und Wissen Volkseigener Verlag, 1961), 21.

37. Pöschel, op. cit., note 16, 29.

38. Stenographic notes of the Second Plenary Conference of the Research Council of the GDR on 12 November 1962 in Berlin. Quoted by Pöschel, op. cit., note 16, 42-43.

39. C. Freeman, The Economics of Industrial Innovation (Harmondsworth: Penguin, 1974), 230.

40. H. Emmerich, "Der Plan 'Neue Technik' erfordert eine höhere Qualität der Leitungstätigkeit", Einheit, Nr. 5 (1961), 698.

41. W. Döring, G. Feldmann and W. Riede, Der Plan Neue Technik. Planteil TOM und der Meister (Berlin: Verlag Die Wirtschaft, 1961), 13.

42. P. Blume, "Der Plan Neue Technik als Voraussetzung zur planmäßigen Erfüllung des Rekonstruktionsplanes", Die Technik, 16, Nr. 10 (1961), 676.

43. Die Aufgaben zur weiteren ökonomischen Stärkung der DDR und zur Festigung der sozialistischen Demokratie (Berlin: Dietz Verlag, 1961), 195.

44. H. Grosse, "Für eine höhere Qualität der Pläne Neue Technik", Die Wirtschaft, Nr. 40 (1960), 7.

45. Döring et al., op. cit., note 41, 26.

46. "Beschluß über die Ausarbeitung und Anwendung von Betriebsprämienordnungen an den volkseigenen und ihnen gleichgestellten Betrieben", GBl, T II, Nr. 14 (1962), 119-120.

47. Ibid., 120-121.

48. O. Kratsch, "Zur sowjetischen Diskussion über 'Plan, Gewinn, Prämie'", Wirtschaftswissenschaft, Nr. 1 (1963), 110. I am indebted to Ulrich Wagner for this and some other sources used in this section. See: U. Wagner, Interessenkonflikte zwischen politischer Führung und Betriebsleitungen in sowjetischer Zentralverwaltungswirtschaften (Dissertation Marburg, 1967).

49. K.H. Jonuscheit, "Das ökonomische System der Planung und Leitung unserer Volkswirtschaft, das Prinzip des demokratischen Zentralismus und die materielle Interessiertheit", Einheit, Nr. 5 (1963), 64-65.

50. C. Gierisch, "Die neuen Bestimmungen über Planung, Vorbereitung, Durchführung und Finanzierung der Investitionen und die Aufgaben der Finanzorgane", Deutsche Finanzwirtschaft, Nr. 3 (1963), 8; "Verschwender von Volkseigentum am Pranger", Deutsche Finanzwirtschaft, Nr. 13 (1962), F9.

51. C. Dewey, "Ökonomischer Nutzen und Planvorbereitung 1963", Deutsche Finanzwirtschaft, 16, Nr. 24 (1962), 9.

52. E. Schwertner and A. Kempke, Zur Wissenschafts- und Hochschulpolitik der SED (1945/46-1966) (Berlin: Dietz Verlag, 1967), 37.

53. Lange, op. cit., note 15, 3.

54. H. Borchert, "Zu einigen Fragen der Planung 'Neue Technik' im VEB Elektrochemischen Kombinat Bitterfeld", Wissenschaftliche Zeitschrift der Martin-Luther-Universität zu Halle-Wittenberg, Ges.-Sprachwiss. Reihe, X1, Nr. 7 (1962), 819.

55. In the early sixties, "commodity production" was also used as main success index. The bonus decree of March 1962, for instance, speaks of "commodity" rather than "gross" production:

GBl, T II, Nr. 14 (1962), 120. "Commodity production" was a superior index since it referred to marketable output. See also Chapter Five.

56. See Chapter Three on the R&D effort; also W. Rumpf, "Staatshaushaltsplan 1963: Steigende Akkumulation - höhere Ausgaben für Produktion und Konsumption", Deutsche Finanzwirtschaft, Nr. 24 (1962), 5.

57. Emmerich, op. cit., note 40, 695.

58. F. Schenk, Magie der Planwirtschaft (Köln/Berlin: Kiepenheuer und Witsch, 1960), 232.

59. See Chapter Two.

60. H. Steeger and G. Schilling, "Die Verbesserung des Systems der ökonomischen Stimulierung der neuen Technik - Ein Beitrag zur Beschleunigung des wissenschaftlich-technischen Fortschritts in der Industrie der DDR", Wissenschaftliche Zeitschrift der Hochschule für Ökonomie, Berlin, Nr. 2 (1961), 107.

61. E. Apel, "Schlußwort zur Diskussion über das Referat und die vorgelegten Entwürfe der Dokumente", in W. Ulbricht, Das Neue Ökonomische System der Planung und Leitung der Volkswirtschaft in der Praxis (Berlin: Dietz Verlag, 1963), 279.

62. H. Fritzsche, "Mehr materiellen Anreiz bei betrieblichen Forschungs- und Entwicklungsarbeiten", Deutsche Finanzwirtschaft, Nr. 2 (1962), F9.

63. H. Steeger, "Der wissenschaftlich-technische Fortschritt in unseren sozialistischen Industriebetrieben und die Sicherung seines höchsten ökonomischen Nutzeffektes", Einheit, Nr. 10 (1962), 20-24; Pöschel and Wikarski, op. cit., note 21, 66.

64. Steeger and Schilling, op. cit., note 60, 111; J. Berliner, The Innovation Decision in Soviet Industry (Cambridge, Mass.: The MIT Press, 1976), 453.

65. G. Hennig and G. Reinecke, "Nutzen und Selbstkostensenkung müssen eins sein", Deutsche Finanzwirtschaft, Nr. 15 (1963), F12.

66. Steeger and Schilling, op. cit., note 60, 106.

67. Friedrich and Schulz, op. cit., note 32, 44.

68. Borchert, op. cit., note 54, 818.

69. Hennig and Reinecke, op. cit., note 65.

70. S. Heyde, "Die Messung der Arbeitsproduktivität und die Praxis", Wirtschaftswissenschaft, Nr. 11 (1962), 1680-1681.

71. Ibid., 1679-1680.

72. "Die Anwendung der fortgeschrittensten Wissenschaft und die Herstellung der Rentabilität in der Industrie", Neues Deutschland, Ausgabe B (19. April 1955), 3.

73. "Bekanntmachung des Beschlusses des Ministerrats über Maßnahmen zur Förderung des wissenschaftlich-technischen Fortschritts in der DDR", GBl, T I, Nr. 63 (1955), 521.

74. Grosse, op. cit., note 44, 7.

75. Rompe, op. cit., note 31, 26. See also Chapter Three.

76. Steenbeck, op. cit., note 14, 681-682.

77. H. Mann, "Probleme der Preisbildung für neue Erzeugnisse (Neue Technik)", Deutsche Finanzwirtschaft, Nr. 23 (1962), 3.

78. "Anordnung über die Finanzierung und Verrechnung der

Forschungs- und Entwicklungsarbeiten in den Betrieben der volkseigenen Wirtschaft", *GBl*, T I, Nr. 82 (1957), 685.

79. "Anordnung Nr. 2 über die Finanzierung und Verrechnung der Forschungs- und Entwicklungsarbeiten in den Betrieben der volkseigenen Wirtschaft vom 28. April 1959", *GBl*, T I, Nr. 32 (1959), 526; "Anordnung Nr. 3 über die Finanzierung und Verrechnung der Forschungs- und Entwicklungsarbeiten in den Betrieben der volkseigenen Wirtschaft vom 21. März 1960", *GBl*, T I, Nr. 22 (1960), 224.

80. S. Schauer, *Ökonomisch begründete Preisrelationen* (Berlin: Verlag Die Wirtschaft, 1965), 32-33.

81. Profit = 6 % of prime costs.

82. See e.g. Mann, *op. cit.*, note 77, 4. For the padding of cost estimates in the USSR see Berliner, *op. cit.*, note 64, 281.

83. See for example the complaint by W. Henschel of the association VVB Industrie- und Spezialbau about inflation of prices for building machines: W. Henschel, "Die Neue Technik darf nicht zu teuer sein", *Die Wirtschaft*, Nr. 22 (1961), 6.

84. C.f. W. Emmrich, "Was die Preisbildung an den Tag bringt", *Die Wirtschaft*, Nr. 12 (1962), 5.

85. A. Dost, "Einige Probleme der Preisbildung zur Erhöhung der ökonomischen Wirksamkeit von Preis und Gewinn", *Die Wirtschaft*, Nr. 44 (1962), 11; Schauer, *op. cit.*, note 80, 35.

86. This could also work the other way around by employing overdimensional standardised components. See Schauer, *op. cit.*, note 80, 146.

87. Schauer, *op. cit.*, note 80, 34.

88. Mann, *op. cit.*, note 77, 4.

89. Schauer, *op. cit.*, note 80, 27-28.

90. "Anordnung über die Gewährung von Gewinnzuschlägen vom 28. April 1959", *GBl*, T I, Nr. 32 (1959), 526-527; "Anordnung Nr. 2 über die Gewährung von Gewinnzuschlägen vom 21. März 1960", *GBl*, T I, Nr. 22 (1960), 223-224.

91. G. Immig, "Gewinnabschläge helfen die neue Technik durchsetzen und werden das Produktionsniveau heben", *Die Wirtschaft*, Nr. 2 (1962), 3.

92. "Anordnung über die Abführung von Gewinnabschlägen zur weiteren Durchsetzung des wissenschaftlich-technischen Fortschritts vom 18. Dezember 1961", *GBl*, T III, Nr. 34 (1961), 399-401.

93. *Ibid.*, 401; see also H. Walther, "Gewinnabschläge für nicht dem technischen Fortschritt entsprechende Erzeugnisse", *Deutsche Finanzwirtschaft*, Nr. 2 (1962), 3-5.

94. *Preispolitik und Preisbildung in der Deutschen Demokratischen Republik* (Berlin: Verlag Die Wirtschaft, 1961), 215.

95. K. Krakat, "Der Weg zur dritten Generation. Die Entwicklung der EDV in der DDR bis zum Beginn der siebziger Jahre", *FS Analysen*, Nr. 7 (1976).

96. W. Kalweit, "Prämien für moderne oder für veraltete Technik?", *Neues Deutschland*, Ausgabe A (15. November 1962), 5.

97. *Protokoll der Verhandlungen des V. Parteitages der Sozialistischen Einheitspartei Deutschlands* (Berlin: Dietz Verlag, 1959), 90-93.

98. See e.g. W. Liebig, "Der Kampf der Betriebspartei-organisation um den wissenschaftlich-technischen Fortschritt", Einheit, Nr. 10 (1962), 136-138.

99. See also the entries in Lexikon der Wirtschaft. Industrie, op. cit., note 2; and DDR Handbuch (Köln: Verlag Wissenschaft und Politik, 1979).

100. J. Hemmerling, "Mit den Neuerungen zum wissenschaftlich-technischen Höchststand", Die Technik, Nr. 11 (1963), 704; Döring et al., op. cit., note 41, 23.

101. Freeman, op. cit., note 39, 242 ff.; and C. Freeman, "Economics of Reserach and Development", Chapter 7 in I. Spiegel-Rösing and D. de Solla-Price (eds.), Science, Technology and Society. A Cross-Disciplinary Perspective (London: Sage, 1977), 257.

102. Kalweit, op. cit., note 96.

5
The New Economic System and Its Successor, 1963–1975

Several factors were instrumental in prompting the political leadership to embark upon a reform of the detailed planning system. Amongst these were the stagnation of the economy at the beginning of the sixties; the freer discussion of ideas from 1956 in the course of destalinisation; the sealing of the border with West Germany; and the need to inspire confidence in a somewhat demoralised population. Perhaps the most important factor was the Liberman discussion in the USSR, which indicated that the Soviet Union was considering an overhaul of its own system. Liberman expounded his ideas in an article entitled "Plan, Profits and Bonuses" published in *Pravda* on 9 September 1962, and proposed a reform of the planning system with "profit" as the main enterprise success index. One month later, Walter Ulbricht spoke of the need to alter the GDR's system, and in January 1963 unveiled proposals for reform which allocated a central role to financial instruments such as profit, prices, bonuses and taxes called "economic levers". The proposals were tested in several enterprises, and in July 1963 the "Principles for the New Economic System of Planning and Management of the National Economy" were approved by the Council of Ministers.[1]

The GDR's New Economic System (NES) was launched two years before that of the USSR, and aimed in particular at reaching "the highest level of science and technology" so as to achieve a "continual increase in labour productivity".[2] This chapter attempts to assess how far the new policy succeeded in overcoming the obstacles to technological change inherent in the system of detailed planning, and what the consequences were when the NES was de facto disbanded in 1971.[3]

BUREAUCRACY AND PLAN COORDINATION

At least three important measures were introduced under the auspices of the New Economic System to reduce the volume of planning activity, and hence alleviate the problems of bureaucracy and of plan coordination. Firstly, there was to be a devolution of decision-making competence. The central agencies were to concentrate on basic targets and long-term goals; the enterprises were vested with more autonomy; and the VVBs were equipped with the status of economic independence. Secondly, certain planning procedures were to be replaced by financial methods. "Economic levers" were to steer enterprise and VVB activity into areas the planners considered desirable for the national economy. Appropriate incentives would, it was assumed, substantially reduce the need for central agencies to exercise pressure for the realisation of technological change. Thirdly, the contents of plans were to be simplified if possible. Whereas the Plan for New Technology in 1962 consisted of twelve component plans, from 1963 it apparently comprised only two; one dealt with the preparation of technological change (the R&D Plan) and the other with its realisation, containing the technical-organisational measures to be accomplished during the plan year (the TOM-Plan).

How effective these measures actually were in relieving problems of bureaucracy and plan coordination is difficult to determine quantitatively, although one approach might be to examine whether the number of important planning indices decreased over the course of the NES, and whether their administration was entrusted to the lower levels of the economic hierarchy. The indices used here are state plan positions (i.e. items in the state plan for important products) and balances (i.e. items balancing supply and demand). Considering first the period 1963-1967, the number of state plan positions fell from 816 to 176: whereas the central agencies (i.e. the State Planning Commission, and National Economic Council or industrial ministries) set all of them in 1963, they were responsible for only 32 % in 1967. Two-thirds of the positions were now set by the VVBs. The total number of balances did not reduce over the period, although there was a considerable devolution: in 1963 the central agencies were responsible for 88 % of the balances and in 1967 they were responsible for only 8 %. Most were now administered by the VVBs.[4]

Some qualifications must be made about this approach, however. For one thing, statistical coverage may vary from year to year and for another, the actual reduction in planning activity was probably less than that suggested by these figures. The central agencies now set fewer indices for the enterprises, but they still demanded an extensive amount of documentation. The economist Uhlemann noted that the enterprise "Bermann-Borsig Berlin" received only six orientation indices and seven "calculation quantities" as state indicators for 1967 but was required to submit approximately fifty separate plan proposals, ranging from the production plan to the scrap plan; the number of plans and documents which the enterprise had to pass on to the higher agencies was in fact larger than that of the previous year.[5] GDR publications indicate that the devolution and simplification of planning was hampered, as in the fifties, by the inertia of the planning agencies. The State Planning Commission apparently continued to include fairly unimportant items in the state plan, an example being the plan position "crates and boxes" in 1965: not only were the enterprises in the packing industry made responsible for fulfilling this position, so too were any manufacturers (such as the "VEB Transformers and X-Ray Works Dresden") which happened to make some of the packing materials for the transport of their products.[6]

The inference for the first four years of the NES is that although the volume of planning activity decreased, there was considerable scope for further reduction, and the devolution and simplification of planning was not so great as that suggested by statistics on state plan positions and balances. The problems of bureaucracy and plan coordination in technological change were alleviated but not overcome. The economist Böttcher stated in 1965 that he had gained the impression, especially through discussions with planning directors, that planning was conducted as an end in itself. Excessive planning, he warned, could transform into its opposite: non-planning.[7] Schmidt and Reinhold of the "VVB Nagema" complained in 1966 about the time and effort needed for the formal paper work required by the various administrative agencies before new technology could be introduced into production. And the Plan for New Technology still lacked internal coordination.[8]

During the period 1968-1970 there was both a partial recentralisation of planning and a greater use of decentralised "economic levers". The amount of detailed planning increased for branches or product groups identified as "structure determining", though no statistical data are available on the number of state plan positions and balances in these years. This increase in planning was part of the leadership's "offensive strategy for technological change", and will be analysed in more detail in Chapter Six. The lower levels of the economic hierarchy, however, received greater freedom with respect to price setting, bonuses and the financing of technological change.

A more or less complete return to detailed central planning took place after 1971. The situation regarding state plan positions in 1972 was virtually the same as in 1963: there were 800 and all were set by the central agencies. Balancing was also recentralised though not to the extent of that in 1963. Delegation was still permitted, but the final responsibility rested with the central agencies.[9] All enterprises were now affected, not just those labelled "structure determining", and the use of "economic levers" was curtailed. The central agencies had, to a considerable degree, recovered their former administrative competence, although as in the period up to 1963, they were poorly situated to know what technological changes were necessary or possible in the enterprises. The problems of bureaucracy and plan coordination for technological change in the post-NES period were probably just as severe as in the period prior to the NES.

The plan for research, development and technical-organisational measures, called from 1969 "The Plan for Science and Technology", threatened to develop into a "mammoth document". In the "VVB Energy Supply", the size of this plan increased as follows: in 1971 it comprised 68 pages; in 1973, 233 pages; and in 1974, 489 pages. Furthermore, other documents were required within the VVB for the management of R&D. The amount of paper work was compounded by the annual changes in planning methodology and by the fact that for many topics the same forms had to be completed each year.[10] From 1972, the VVB had found it necessary to employ fifteen persons full time to deal with documentation for the central agencies. According to G. Nitze of this VVB and H. Beckmann of the Engineering University Zittau, it was difficult enough for "specialists" in science and technology to understand the planning process; for management and workers, it was even worse.[11]

Criticisms about the coordination of planning in technological change made in the seventies were similar to those made in the late fifties and early sixties. The Plan for Science and Technology allegedly lacked both internal coordination, and external relation with other plans. The journal Die Wirtschaft stated in 1974, for instance, that the plan was an "extensive catalogue" of tasks and warned there was a danger of dissipation and duplication of effort unless activities were coordinated.[12] Three difficulties in the relationship of the Plan for Science and Technology with other enterprise plans may be noted. One stemmed from the fact that research activity was best served by flexible planning whereas enterprise management was most interested in the stability of its total planning activity. W. Schwarzbach of the "VEB Pipe Combine/Steel and Rolling Mill Riesa" complained that in the combine's main enterprise about sixty to seventy changes were made to the Plan for Science and Technology each year. These changes had to be confirmed by several administrative agencies, and each change involved a minimum of ten hours work. Schwarzbach felt that science and technology ought to be planned with the same determination as production and recommended that changes in the Plan for Science and Technology no longer be allowed in this manner; that deviations from the original plan target be accounted for; and that the dependability and binding character of the plan targets be raised.[13] A second difficulty was that some enterprises included topics in the Plan for Science and Technology despite inadequate or even non-existent cooperation agreements with other enterprises in the hope that "somehow" the problems would sort themselves out. This brought "confusion to plan fulfilment, often negatively affected wide areas of the national economy", and resulted in "anger and trouble".[14] A third difficulty concerned the synchronisation of R&D planning with investment planning. Investments were planned on an annual basis and the enterprise or combine director had to decide whether to start a particular R&D project, even though he did not know whether he would later have enough investment funds to introduce the results into production.[15]

Two important attempts were made to relieve the difficulties of plan coordination. One was the introduction of a special section "transfer" into the annual enterprise plans for science and technology in 1973/74. This aimed at stipulating and harmonising the measures necessary for successful transfer of scientific results into production.[16] The "Transfer Plan" was hailed as an important improvement to the Plan for Science and Technology, although it is not apparent from the information available how it differed from the earlier plans for including new products and processes in production.[17] The other attempt involved the introduction of "Coordination Plans" for the state science and technology plan. These were adopted from the Soviet Union, tested in the GDR in 1974, and to be introduced in the five-year period 1976-1980.[18] It is extremely doubtful, however, whether such methods of relieving problems of planning through more planning could be very successful. Indeed it is possible they created more coordination difficulties than they solved. It was reported of the Transfer Plan in 1975, for example, that it was only worked out for one year at a time and did not guarantee the longer-term continuity of the project. The tasks of the central Transfer Plan were apparently not properly synchronised with those of the enterprise, and the planning of transfer tasks, research, investment and production was not "comprehensive" enough.[19]

PROJECT SELECTION

Under the system of detailed central planning, research projects were frequently selected according to the purely internal interests of the institute, the producer enterprise and especially its R&D establishment. The central agencies had little more than a statistical overview of research topics, "defence" procedures did not function satisfactorily, and enterprises had scant incentive to introduce major new developments into production. At most, the price and bonus systems directed the enterprise to encourage and implement small cost-saving projects once the price of the product had been set.

The New Economic System introduced financial instruments which were intended to provide the centre with a better overview of major research projects, and to stimulate the interest of enterprise management and R&D workers in projects which would result in significant technological change and the satisfaction of user needs. Important instruments in this respect were decentralised R&D financing and modified price and bonus systems. Probably the most far-reaching changes were introduced during the period 1968-1971, and in this section the influence on project selection, before and after 1968, of decentralised R&D financing and of bonuses for technological change will be considered. The stimulus to technological change provided by the price structure and the general enterprise bonus system will be examined in the next two sections.

Most industrial R&D prior to the NES had been financed in the first instance from the state budget. The manufacturer was supposed to pay back this finance by including R&D costs in the prime costs of the product.[20] In practice repayments did not match state expenditure, and if an R&D establishment discontinued a project there were usually no financial repercussions: the costs were written off. In 1963 a "Fund for Technology" was established at VVB level, and was to work according to the principle of "economic accounting", i.e. that expenditure be covered by income. It was based on contributions from the enterprises and was to finance:

- R&D work, including prototypes, pilot lots, experimental plants, and oriented basic research;
- Contract R&D conducted outside the VVB;
- Start-up costs resulting from the introduction of R&D results into production;
- Fixed assets, equipment, tools and training, provided these were necessary for the R&D work;
- Costs for GDR- and special standards;
- Samples for comparison with the "world level" of technology;
- Licenses;
- Bonuses for contract research. Bonuses for other R&D work were financed from the general bonus fund (VVB and enterprise level) and from so-called "disposal fund" of the VVB general director.[21]

Now only tasks especially important for the development of the leading sectors of the economy would receive finance from the state budget.[22] The financing of R&D was thus more decentralised, though it should be noted that a certain amount of centralisation from enterprise to VVB had also taken place: the rather limited enterprise "Fund for New Technology" created in 1959 was now redundant.[23]

In principle, the new method of funding R&D had several important advantages. One was that the centre now had a better chance of identifying and supporting projects likely to result in significant technological change because it was no longer overburdened with details of all current and proposed projects. Another advantage was that since the VVBs possessed more detailed knowledge than the centre of enterprise technology and now also administered most of the institutes, they could prevent project choice being determined too greatly by the personal inclinations of individual R&D workers or the interest of enterprise management in small, riskless, cost-saving projects. Both the centre and the VVBs attached greater importance than before to the determination of economic effectiveness as an aid to project selection. Two further advantages of the fund were that it relieved the enterprise of the problem of procuring finance for the start-up, and helped keep the price of the new product low, since research, development and start-up costs were now, in effect, included in the costs of all the VVB's products and not just in the costs of new products.

As with the devolution of planning in general, it is difficult to judge the extent to which in practice competence for R&D financing was conferred to the VVB. It is also difficult to determine whether the centre and the VVBs actually played a more effective role in encouraging technological change. Indications that the centre was not greatly successful in selecting and supporting projects promising significant technological change were provided in 1966 by Pöschel, of the department for R&D attached to the Central Committee, and in 1968 by Kusicka, an economist specialised in questions of R&D: important projects were supposed to be included in the State Plan for New Technology, ones of lesser importance appearing in the enterprise plan only, and apparently many of the tasks contained in the State Plan were not even planned at the highest scientific-technical level. Accordingly the results were of only a mediocre standard.[24] Indications that the VVBs did not play a very effective role were provided by continued complaints about R&D "defence" procedures and the lack of orientation to user and national needs. Calculations of economic usefulness did not provide a short cut to effective project choice. The costs of a project and the likely benefits could be predicted only with difficulty, and the answers received depended heavily upon the calculatory methods used. In 1965 a "Draft of a Methodology for Determining the Economic Usefulness of Tasks in Scientific-Technical Progress" was published.[25] This was criticised for requiring such a large number of investigations that it could not be used in practice.[26] As in the West, the calculation of "economic effectiveness" was often regarded as a formal exercise to justify carrying out a particular project.[27] In many cases it was first calculated on completion of the project.[28]

Clearly the greater the stimulus to technological change provided by the bonus and price systems, the less direction was necessary from the centre and VVB. Unfortunately, the bonuses specifically intended to reward research, development and transfer of results into production were not very effective in this respect. Such bonuses were still awarded schematically, were based too much on age and position and too little on the utility of the results and the individual performance of the R&D worker.[29] The rules, according to Kusicka, were extremely complicated and unclear,[30] and it is likely that many enterprises still linked R&D bonus payments to the overall fulfilment of the main annual enterprise plan. This would help explain why enterprise R&D collectives tended to concentrate their attention onto part two of the Plan for New Technology, i.e. the annual TOM-Plan, and within this plan onto such topics as would help the enterprise attain its main plan targets for that year.[31] R&D collectives were not inclined to include projects in the TOM-Plan which would result in a comprehensive improvement of enterprise technology, for the higher the level of enterprise technology, the more difficult it was to fulfil this plan.

The research institutes were less oriented to the annual enterprise plan but even so the "end-of-year bonus" was a common type of bonus. Investigations in the ten institutes of the "VVB Machine and Vehicle Building" revealed, for instance, that this bonus accounted for about 40 % of total bonus payments in 1966 and for 74 % in 1967.[32] Some critics argued that the end-of-year bonus was inappropriate for research and development. E. Dreyer, economic director at the Institute for Data Processing, stated that his institute had completed a number of R&D topics in the first quarter of 1967. If end-of-year bonuses had been awarded in December 1966 or January 1967, there would have been no financial incentive to conclude the projects successfully. If such bonuses had been awarded at the end of 1967, too long a period would have elapsed, and again the bonuses would have lacked effectiveness.[33] Others stressed that R&D projects frequently stretched over a number of years and involved different persons at different times: bonuses were best awarded on an annual basis for the successful completion of controllable intermediate steps. Furthermore, "the orientation of our party and government towards the end-of-year bonus is unequivocal".[34] In the investigations conducted in the institutes of the "VVB Machine and Vehicle Building" it was found that the end-of-year bonuses were not based on utility and individual performance but on wage classification and the length of time worked.[35] So-called "target bonuses" which could be awarded for reaching a particular target in R&D work, and were not confined to the planning year, accounted for about 20 % of the total bonus fund in 1966. Instead of rewarding the excellence of the results, however, they were usually granted for the completion of a research task before a certain date.[36] The motivation then was to overstate the time needed for completing an R&D task, and to avoid work involving too much risk.

At best, bonuses for technological change up to 1968 were neutral with regard to stimulating significant, user-related technological change. At worst, they directed research and technical effort towards relatively unimportant, easily fulfilled tasks.

From January 1969 there was a further decentralisation in the financing of R&D. Now not only VVBs but enterprises were equipped with a "Fund for Science and Technology" which they had to maintain themselves. The fund was financed via a cost component in product price, together with income derived from contract work and the sales of licenses, trial production and fixed assets.[37] As most R&D tasks had to be conducted henceforth on a contract basis, R&D personnel were given an incentive to concern themselves with projects relevant to the user. According to the Minister for Science and Technology, Günter Prey, the Fund for Science and Technology at VVB level was in future only to contain finance necessary for "structure-determining tasks, basic research and other cross-sectional problems".[38] This fund was built from contributions from the enterprises' Fund for Science and Technology.[39] If a proposed R&D project was considered very important for the economy, i.e. was a "structure-determining task", but was beyond the financial means of an enterprise, combine, or VVB, then funds could still be obtained from the state budget.[40]

The effectiveness of the new method of financing R&D depended in practice upon at least two important factors. One was how far the price and bonus systems actually succeeded in stimulating enterprises to pursue technological change. Another was how far the higher authorities were prepared to allow a devolution of financing competence; critical in this respect was the interpretation of the term "structure-determining areas". If the authorities interpreted the term in a broad sense, then the new system in practice would not have differed greatly from the old one: a substantial amount of finance would still have been provided from the state budget, and the VVBs would have centralised finance for technological change in their own, rather than the enterprises', Funds for Science and Technology.[41] On the other hand, if the term was interpreted in a narrow sense, R&D establishments would have been prompted to seek "safe contracts". When the NES effectively came to an end in 1971, the enterprise Fund for Science and Technology was not abolished but it is likely that the higher authorities reassumed much of the responsibility for financing R&D.

The Fund for Science and Technology was not intended to finance bonuses for technological change. In 1969 two new funds were created in R&D establishments, aimed at increasing the effectiveness of such bonuses: the bonus fund for R&D and the "performance fund". The first was financed from, and usually amounted to 25 % of, the second.[42] The "performance fund" was formed from income derived from scientific-technical work, in particular a "performance-related surcharge" included in the price of R&D work. The customer paid the surcharge if the work was completed within the period agreed and was of acceptable quality. For projects important to the national economy it amounted to between 20 and 40 % of the contractually agreed wage costs. For the remaining projects in the Plan for Science and Technology, it was between 10 and 25 %. Other types of scientific-technical projects were entitled to a price surcharge of between 5 and 10 %. [43] If the "performance fund" was too low to finance even the minimum level of the bonus fund, the administrative agencies could arrange this payment from other sources. [44] In 1970 the engineering bureaux, concerned with rationalisation measures in the VVBs, went one step further: from now on their price surcharge was to depend upon the economic benefit to the user. [45] In spite of these changes, there continued to be criticisms in the period up to 1975 about schematism in the award of bonuses for technological change, and their insufficient relation to utility and individual performance.[46] So-called "household books" were introduced to document individual performance, but in practice they were often used simply to record costs. [47]

GDR publications in the post-NES period indicate that the authorities were dissatisfied with the effectiveness of project selection. Two economists, Marschall and Lange, cited the findings of investigations into obstacles in the GDR's "research-production cycle". Deficiencies in project selection represented the most important obstacle, accounting for 42 % of the economic loss; "complications in research" accounted for 23 %; and inadequacies in the "transfer" for 35 %. [48] Such statistics must be viewed with some caution, however: with hindsight it may well be apparent that the choice of a particular R&D project was a mistake, yet this hindsight was gained by actually conducting the research; the knowledge was probably not available at the time of project selection. Perhaps more telling than such statistics were the criticisms relating to project selection reminiscent of the late fifties and early sixties. The central Plan for Science and Technology was still a compilation of topics from below;[49] projects still took inadequate account of utility; market research was weak; and "defence" procedures remained internal affairs and little more than a formality. [50] In fact, first contact between an R&D collective and its "partners" sometimes took place in the State Contract Court. [51] Some managers of enterprises and combines still failed to involve themselves with the planning of science and technology, and consequently project choice often continued to depend upon the personal interests of individual scientists and engineers, to the neglect of the needs of the user and the national economy. [52]

PRICES

During the course of the New Economic System, measures were adopted to make prices more flexible and in particular, more dependent on quality and the benefit to the user. In 1964 a system of price surcharges and discounts was introduced to promote the manufacture of high quality products. It applied to products of the metalworking industry, manufactured by state enterprises and classified by the German Bureau for Measurements and Quality Control (GBMQ). Products of "world-market" quality having the highest state classification "Q", were permitted a surcharge equivalent to 2 % of the enterprise price. Products in the next state category "1" had no surcharge, but those in the lowest category "2" were imposed a price discount of 5 %.[53] The method of adjusting enterprise profit introduced from 1959 for new and old products still remained in force for industry as a whole,[54] though provision was made to apply only the price adjustment system for new products of quality "Q" and old products of quality "2" in the metalworking sector.

Price adjustment had several advantages over enterprise profit adjustment, especially with respect to new products. Firstly, the price surcharge was more oriented towards rewarding quality and utility since it was not sufficient for a product simply to be "new" in order to be granted a supplement: it had to reach the level "Q". Secondly, its period of operation was as long as the product carried the seal "Q" and not restricted to one year. Thirdly, the enterprise did not suffer as greatly from the deterioration of auxiliary indices on introducing new technology. An advantage of the price discount over the profit deduction was that the GBMQ did not have the VVB's vested interest in demonstrating "profitable" production. In spite of these advantages, the surcharge and discount were usually not large enough to motivate enterprises to introduce new or improved technology. This was especially the case for enterprises with small-series production.[55]

The system of price adjustment also had some disadvantages compared with the adjustment of enterprise profit. Firstly, it was limited in scope as it applied only to the metalworking sector and then to those products that had to be tested and classified by the GBMQ. For machine building in 1965 this amounted to 60 % of the products, and within the various branches of machine building the proportion of products to be tested varied greatly: in food machine building, for instance, it was only 35 %.

Since most tested products had the seal "1", the price surcharge and discount came into question for only a small proportion of the products. Secondly, the new system was probably more bureaucratic than the adjustment of enterprise profit. Admittedly enterprises were now entrusted with the task of changing the prices themselves with regard to addition of the surcharge or subtraction of the discount.[56] They could only do this, however, after the GBMQ had classified the product, and this apparently took considerable time. A third disadvantage of the new system was that the price discounts may have stimulated demand for old products rather than phasing them out.

From 1967 a new method of price adjustment was introduced to promote the manufacture of new and improved products and the exclusion of old ones. It had several advantages over the previous method. The surcharge was based on the economic usefulness of the product to the user, and was incorporated into a degenerating price curve which was intended to overcome the increasing profitability of products with time. The discount penalised enterprises manufacturing old products, but could avoid stimulating additional demand. Price setting under this method was decidedly more decentralised and not ultimately dependent on the GBMQ. The system of degressive prices was introduced not only into the metalworking industry, but also into the chemical industry in 1968 and, to promote the use of new and improved processes, into the foundry industry in 1969.[57] Price adjustment according to the quality classification system still remained in force for the metalworking industry, but for new or improved products of quality "Q" a surcharge was applied according to the new system if it exceeded the usual 2 % price mark-up.[58] The method of enterprise profit adjustment, however, could no longer be applied in the three sectors of industry using the new price adjustment method.[59]

The surcharge reaped by the producer could be up to 30 % of the economic use to the main customers, but could not exceed twice the calculatory profit mark-up.[60] The procedure was termed "benefit sharing". The degression was first agreed upon by the producer enterprise and its main users, and then confirmed by the VVB.[61] It was supposed to be such that the maximum price was allocated to a new or improved product at the "highest scientific-technical level". The price was then to decrease over the economic life of the product and when the product was no longer at the highest scientific-technical level, the price was to fall faster than the prime costs. In this way the manufacturer would have a strong incentive to cease production. Such an incentive was necessary, for if the introduction of new products into manufacture had been slow, the removal of old ones had been even slower: in 1965, roughly twice as many products were introduced into manufacture as were removed, and in the opinion of one GDR author, the product range in GDR industry had become swollen to an "economically unjustified" degree.[62] Since the degression was fixed in advance, the changing price would not lead to loss of information to the higher economic agencies nor to confusion in enterprise plan fulfilment. In the event of an incorrect estimation of "moral depreciation" (i.e. obsolescence), user enterprises, the GBMQ, and the banks had the right to apply for a correction of the degression.[63] The discount could avoid stimulating demand for old products through a provision enabling the VVB to cut the profit for the producer but maintain the old price for the customer. The difference was then paid into the VVB's Fund for Technology.[64]

The system of degressive prices was an imaginative advance in GDR price formation, although its use in practice was not entirely satisfactory. Two general problems may be mentioned before discussing difficulties specific to the surcharge, degeneration curve and discount. One problem stemmed from the role assigned to the VVBs. These were responsible for administering the prices but since they were concerned to show a respectable level of profitability, it is difficult to discern what interest they had in ensuring that initial prices were not inflated and that their enterprises actually applied the degression and discount. Criticism of producer prices was supposed to come from the users, though this was not very well developed.[65] An important reason for this was that users were not sufficiently informed about the "reserves" of producer enterprises. The new price method in fact made it easier for producers to drive up prices through minor product changes. Such simulated technological change was also encouraged by the official propaganda of the period 1967/68-1971 which stressed the aim of overtaking West Germany through new technology. A second problem concerned the stipulation that the new price system was not to be applied to products which were also sold as consumer goods.[66] This meant that even if a very small proportion of the output of a certain product was used as consumer goods, the whole of this output was excluded from the system of degressive prices. It was the reason why in the "VVB Electrical Appliances" the new price system was applied to only about 15 % of the products.[67]

Concerning the surcharge, one drawback was that it was limited to twice the calculatory profit mark-up. This did not adequately reward enterprises for bringing forth new or improved products yielding a high economic benefit. The situation was particularly severe for final producers having predominantly one-off and small-series manufacture; they played an important role in developing new technology but made only a small profit.[68] In practice, many of the prices for new products in the metalworking industry did not include a surcharge.[69] Economic benefit depended on where and how the new product was utilised, and what parameters were used to measure it. Producer enterprises were inclined to exaggerate the utility of a new product whereas user enterprises had an interest in underestimating the economic benefit - a high estimation would have meant higher prices for them. Investigations showed that it sometimes took several months before agreement could be reached between the producer and the user about the level of the surcharge. Many users in fact failed to provide producers with information on economic usefulness or were reluctant to cooperate in its calculation. Consequently a number of producers did not press for a price surcharge and instead ensured that cost estimates for the basic price were sufficiently "padded". Some producers even maintained that if the price contained no surcharge, then users could not demand a degression.[70] In an attempt to enforce the method of surcharges, an order was passed in May 1970 prescribing that in future, surcharges should be included in the price calculation and be based on economic usefulness. To circumvent too long a dispute between the manufacturer and user about the size of the surcharge, the price agencies were empowered to dictate its value if necessary.[71]

With regard to fixing the stepped degression curve, no-one had detailed knowledge of the potential use of a product and the future development of prime costs. An "incorrect" degression would lead either to prime costs falling more rapidly than the price degression so that the enterprise had an interest in postponing the planned withdrawal from manufacture, or to the product being taken out of manufacture before the end of its economic life cycle owing to the small or negative profit. In practice the step-function curve was often dramatically simplified. A common method was to specify when the surcharge was to be reduced and removed, but not to adjust the price further in view of the lower prime costs. The "VVB Diesel Motors, Pumps and Compressors" set its introductory prices for two years, to be followed by the price without surcharge for five years. The "VVB Electrical Appliances" had a three stage degression curve: the initial price; the removal of most of the surcharge; and finally the removal of the remaining part of the surcharge together with half of the calculatory profit.[72] Clearly, the simplification of the step-function meant that the problem of the increasing profitability of old products was not overcome. A positive aspect of this practice with regard to the diffusion of new or improved technology, however, was that user enterprises were not inclined to delay buying a product in the knowledge that the price would be at a lower stage of degression later on. To force enterprises to employ a price degression which not only cut and removed the surcharge, but also reduced the basic price of the product faster than the prime costs, the order of May 1970 prescribed that by 31 December of that year, industrial prices for all products of the metalworking industry affected by the original decree of 1967 had to be structured in a degressive manner. The only exceptions were standardised products and products having a long economic life. In these cases the degression was limited to the removal of the surcharge only.[73]

The effectiveness of the price discount depended on whether products were declared old either by the producer enterprise itself, its VVB, the GBMQ, or the main users. Clearly the producer had little financial interest in having its own products pronounced old. The VVB would have been at least as reluctant to classify too many of its enterprises' products old as it was under the 1961/62 system to impose too many profit deductions. The GBMQ was not sufficiently well informed about all enterprise products to impose price discounts. And even the users were not very active in criticising old products, probably because they feared price increases on the introduction of new ones. As with the 1960/61 method of enterprise profit deductions, there was a loophole in the law which permitted the enterprise to avoid applying price discounts. According to one paragraph, discounts could not be applied to old products, if these products were required domestically or for exports.[74] Since new or further developed products often could not be manufactured immediately on a large enough scale, producer enterprises were able to use this loophole to continue manufacturing out-of-date products at the old price.

Finally, one further general problem of the system of degressive prices should be noted. This concerned its lack of coordination with the other "economic levers" of the NES. There was apparently confusion about how the surcharges, degressions and discounts were to be considered in plan fulfilment and the award of bonuses. There was also confusion about how the degressive price fitted together with the other types of flexible price introduced from 1968 such as the "maximum price", the "agreed price" and the declining capital-based price. Fixed prices were to be restricted primarily to products of fairly constant use-value, particularly in the basic materials industry. Most products were to have "maximum prices". These prices could be changed quickly to adapt to demand conditions without involving the participation of the superior agencies, but could not be exceeded. This was an important improvement in price formation, though in practice their use was probably impaired by the existence of a seller's market. The "agreed price" was prescribed for special or single orders, complete plants and R&D work. It was agreed upon by the supplier and customer on the basis of "benefit sharing", but no fixed profit margin was prescribed, and the price was not confirmed by the price agencies.[75] The weakness of this type of price was that if only one supplier came into question for a particular product, the customer had little option but to pay the price demanded.

The declining capital-based price was to be introduced into four VVBs and combines in 1969 and into twenty-four from 1970. The price aimed at encouraging a more effective use of capital, and at overcoming the increasing profitability with time of older products. The first aim was to be achieved by the imposition of a "production fund charge" on the enterprise, payable out of profits. The second aim was to be realised by prescribing declining upper and lower limits to the profit of a product group per 1,000 Marks of necessary fixed and working capital. On exceeding the upper limit, a price reduction to the lower limit had to be introduced from the beginning of the following year.[76] This meant, of course, that the profitability of individual enterprises may have lain outside the limits prescribed for the groups. High-profit enterprises would have had little incentive to improve on their costs, whereas low-profit enterprises may have experienced some hardship. The group as a whole had scant interest in approaching the upper profit limit, for a price reduction would have made it all the more difficult to earn bonuses in the future; instead it was more attractive to bring about price increases through minor product changes.

The year 1971 saw many of the changes towards a more flexible price system put into reverse. The general recentralisation of planning was accompanied by a price freeze up to 1975 (and later up to 1980).[77] Interestingly, price degression was broken off in the GDR in the same year that it was introduced into the Soviet Union.[78] Price confirmation was now exceedingly bureaucratic. No longer were the VVBs permitted to administer prices; they could only check price applications from the enterprises and then submit the documents to the higher agencies. For important new or improved products, prices had to be confirmed by the Council of Ministers, the minister and director of the Price Bureau and the relevant industrial minister. Other products had to be confirmed by the Price Bureau.[79] This agency was charged with the task of ensuring "price discipline as an important component of socialist state and plan discipline".[80] It was empowered to impose fines on enterprises, combines, VVBs and ministries; confiscate unjustified profits; and shame offenders in public.

Despite these developments, the price rules of the seventies at least in principle offered more stimulus to technological change than had those prior to 1963. Surcharges and discounts were, as in the NES, applied to price rather than to enterprise profit. The price surcharge was based on economic usefulness, unlike the enterprise profit supplement, and its period of operation was three years rather than one.[81] The 1967 principle of "benefit sharing" was replaced by the level of the surcharge being set by the Price Bureau. Users were obliged to provide information about the usefulness of the product. As in the sixties, a product classified by the GBMQ was granted a surcharge if it was judged to be of level "Q". The new rules also permitted products of level "1" a surcharge if a higher economic effect for the user could be demonstrated. New products which did not differ very much from the ones they were due to replace in manufacture were, however, not granted a higher price.[82] The rules for surcharges and discounts of the seventies differed from those of the NES, but the basic problems remained the same. Concerning the surcharge, for example, the calculation of economic usefulness was difficult, and user enterprises had little interest in facilitating it. Producer enterprises did not want to declare their reserves. Nor were they very interested in receiving an additional profit for three years, as after this period they could experience difficulty in annual plan fulfilment.[83]

In practice, the price rules in the seventies offered little more stimulus to technological change than those of the detailed planning system. The old problems of fixed, cost prices remained. A typical complaint was made in 1973 by N. Scholz, technical director of the machine-building factory "Heidenau" in the "Combine Nagema":

> We are currently working on a new product which ... will surpass its predecessor in capacity by 20 % and be about 25 % less material intensive. Moreover, there are a number of other advantages for the user. Since it involves a new labour principle, we will not be able to adopt the price of the old product but instead will have to calculate it anew and reckon with a lower price. However, as we can then no longer achieve the planned commodity production, we really ought to continue manufacturing the old product in 1974.[84]

The economists Mann, Banse, Nick and others pointed out that it was not in the interests of the enterprise to implement optimal project proposals for new products. If use-value/cost analysis were rigorously applied, this could preclude the possibility of reducing prime costs in future plan periods and hence jeopardise plan fulfilment. A better strategy from the point of view of the enterprise was to keep possible product improvements in reserve.[85]

Up to 1975 relation prices were still not used extensively for new products, despite the fact that they were not so arbitrary as cost prices and did not necessarily penalise the manufacturer for low costs. Theoreticians were still wrestling with the problem of how to reconcile Marx's labour theory of value with the practical necessity of incorporating utility considerations in prices. Many felt that although use-value could enter price considerations by a deliberate deviation of price from value (as in the case of the cost price and surcharge) it could not in itself be a constitutive element of value determination. Thus with parametric prices, use-value was only supposed to enter price formation indirectly via producer parameters, i.e. those reflecting the "societally necessary labour expenditure", and not directly through user parameters.

> If this is not guaranteed, the price, as a result of the continual improvement of the product's use characteristics, may depart from its objective basis of determination, the societally necessary labour expenditure, and industrial prices can be unjustifiably set at too high a level. If use-value were recognised as a price-determining factor, the law of value would be infringed. Value would then be practically exchangeable

> for use-value. This contradicts the Marxist theory of labour value and leads in the final analysis to a subjective doctrine of value.[86]

Not all economists agreed with this however. Banse and Nick, for instance, argued in 1975 for prices of new products to be formed in relation to the use-value of comparable products. Giving authority to their arguments by selected quotations from Marx and from Soviet writers, they urged that prices be established according to a "price-performance relationship" rather than individual enterprise costs.[87] The pragmatists succeeded in convincing the political leadership, for in 1976 a decree was published prescribing price formation along these lines for new and improved products.[88]

THE ENTERPRISE BONUS SYSTEM

The general enterprise bonus system in force up to 1963 provided little incentive to engage in technological change, three aspects of the system being mainly responsible: the measurement of enterprise success relative to the plan; an operative plan period of only one year; and the use of "gross production" as the main success index. During the course of the NES, modifications were made to the bonus system in an attempt to overcome these difficulties.

The first modification occurred in 1963: the main success index was now "profit" rather than "gross production".[89] Profit was allocated a central role in the NES and was intended to stimulate economic efficiency and technological change, and to serve as a source of finance for investments.[90] The old success index was retained as a subsidiary indicator, though in somewhat superior form. Now called "commodity production", it referred to marketable output and excluded such things as internally used production; the cost of meeting guarantee obligations; and the value of products registered under the State Classification System which possessed no quality grade and which should already have been withdrawn from manufacture. Also excluded was the cost of R&D work that had not involved the manufacture and marketing of prototypes, pilot lots and large-scale testing models.[91]

Prior to 1968/69 it is unlikely that the new main success index provided much more stimulus to technological change than had "gross production". Enterprises were now interested in manufacturing those products with a high profit margin, and these were generally not new products. As noted in the previous section, even with the price surcharge introduced in 1964, new products were not very attractive for the manufacturer. To counteract this tendency, the National Economic Council toyed in 1965 with the idea of withdrawing profit as the main index for 1966.[92] Keren states in a couple of publications that profit was abolished in 1965 as the main success indicator "for a year or so",[93] yet scrutiny of the laws for the 1965, 1966 and 1967 bonus funds does not support this assertion.[94] In addition to the difficulty of the profit index, enterprise success continued to be measured relative to the plan and the operative plan period was still only one year. The planners were interested, as before, in "taut" enterprise plans and the "ratchet"; the "tautness" of the aggregate target for industrial output rose from a planned 5.4 % increase in 1964 to one of 6.4 % in 1968.[95] Enterprises still fought shy of implementing significant technological change and continued to be interested in small, cost-saving changes once the price had been set, for this increased the margin between planned and actual profit and hence resulted in higher bonuses. They were careful not to implement too many small changes in any one year, for this might have attracted a higher profit target in the succeeding year and perhaps a lower product price. Prudent enterprises would still have pursued "slack" plans for technological change, particularly when they had "state tasks" in their Plans for New Technology as the fulfilment of these tasks was made a condition for the full payment of enterprise bonuses.[96]

Interestingly the 1964 law on enterprise bonuses included provisions to overcome enterprise interest in a "slack" main plan index and promote what the planners considered to be a more rational economic behaviour. The method was to give the VVB and its enterprise the opportunity of earning high bonuses by "overbidding" the orientation profit index proposed by the State Planning Commission. The VVB was permitted to plan an extra contribution to the bonus fund equivalent to up to 75 % of the "overbid" profit. Providing this target was fulfilled, the VVB head office was to distribute the extra bonus contribution amongst its enterprises according to their performance, after deducting up to 7 % for itself. If the overbid-target was not fulfilled, the VVB received in 1964 up to 40 % of the difference between actual profit and orientation index. In 1965 the proportion was to be up to 20 %. The VVB and its enterprises could choose not to "overbid" and simply overfulfil the profit index. In this case enterprises in 1964 were allowed to channel up to 60 % of their extra profit into their bonus funds. In 1965 the limit was 30 %.[97]

This law had two major weaknesses. Firstly, its objective of promoting "optimal plans" could not be realised, for the measurement of success was still made relative to the plan and it remained in the interest of the enterprise to keep targets low. This meant either not "overbidding" or only with moderation; the enterprise was then better equipped to deal with any future "ratchet effect". As one GDR economist pointed out, many economic functionaries were more afraid of reprimands from the state apparatus and the party for not fulfilling the plan than they were attracted to extra bonuses for "overbidding" and proposing "optimal" plans.[98] A second weakness of the law was that it had conflicting objectives: it simultaneously aimed at promoting "optimal plans" and at promoting technological change, especially by including fulfilment of the Plan for New Technology or the Export Plan as subsidiary indices.[99] The more "optimal" the plan, the less likelihood there was of new technology being introduced. The engineer Wesselburg commented in 1965 on the index profit, slack plans and avoidance of technological change as follows:

> The plan index profit is also reset annually and a hard battle is fought out between enterprise and central planning respectively. The uncertainty still inherent in plan estimation, and thus the plan-target profit, stimulates the enterprise to seek a certain security and keep the plan-target profit low. A general phenomenon of our economy is the aversion of enterprises and economic branches to accept additional or new production. Often, debates lasting weeks and months develop between the enterprises and VVB, usually with the inclusion of the National Economic Council, about who should take on a necessary new production ...[100]

As noted in the previous section, steps were taken from 1967 to make new products more profitable than old. In 1968 measures were introduced to deal with the problems caused by relative plan fulfilment and the short operative plan period: enterprise success was to be measured in absolute terms and over longer periods than one year. Perspective plans, in the first instance from 1969-1970 and 1971-1975, were now to be of greater importance than annual plans. Enterprises were permitted to carry over bonus funds from one year to the next within a perspective plan period, or to use part of the fund in advance.[101] No planned profit target was stipulated for the enterprises: instead bonuses were to depend on the increase in net profit, i.e. on the increase of profit after deduction of the production fund charge. The bonus fund for 1969 was established according to the following formula:[102]

Bonus Fund =	"Basic Contribution"	+	"Fund Increase"
=	x % of planned net profit for 1968	+	y % of increase in net profit in 1969 over the planned net profit for 1968

The "normatives" x and y were specified by the ministries for the VVBs, and by the VVBs in turn for their enterprises. To support the party's structural policy, higher proportions were to be allocated to enterprises classified as "structure determining". In 1970 the "basic contribution" was to be raised by an amount equivalent to 15 % of the 1969 "fund increase". The "fund increase" for 1970 was to be based on the increase of actual net profit between 1969 and 1970.[103]

The new bonus rules reduced the problems of taut plans and the ratchet effect but did not surmount them. A ratchet effect would probably have occurred between the perspective periods (i.e. in 1970, 1975 etc.). The VVB was concerned to show a reasonable rate of profitability and on receiving normatives from the ministry, it had to set normatives for the bonus funds of its enterprises, together with the levels of a differentiated charge on enterprise net profit to be paid to the state. If an enterprise made a high net profit in one perspective plan period, its VVB would have been inclined to set lower normatives and a higher differentiated charge for the next plan period. In this way the VVB could have compensated for the performance of less profitable enterprises. An enterprise would also have been unwise to engage in significant technological change towards the end of a perspective period under these bonus conditions, as again the VVB may have set lower normatives and a higher differentiated charge for the next perspective period in expectation of higher profits.

Enterprises continued to be afraid of taut plans for at least three reasons. One was that although "gross production" was no longer the main enterprise success index, the party leadership was still concerned that industrial output targets of the national economic plan be fulfilled. The method adopted was to incorporate planned enterprise output into contracts, the violation of which led to indemnities. Another reason for fearing taut plans was that even though profit was measured in terms of its increase over the previous year, the subsidiary indices were still measured relative to the plan. To qualify for full bonuses, enterprises had to fulfil two state plan indices stipulated by the superior agencies from the following list:[104]

- Export tasks;
- Tasks in science and technology;
- "Structure- and proportion-determining deliveries" to important users in the domestic market;
- Deliveries and services for "structure-determining investments";
- Sale of finished products for the population;
- Increase in the output-capital ratio, especially through the optimal use of highly productive machines and plants.

Non-fulfilment of these indices incurred much heavier penalties than had non-fulfilment of auxiliary indices in the past.

According to Rudolf Wiesner of the State Bureau for Labour and Wages, more than two auxiliary indices had been used previously, but if they were not fulfilled the bonus fund was reduced by only a small amount. The degree of reduction had in fact been scaled to the degree of non-fulfilment of the plan and in consequence, enterprises had not worried about fulfilment of the auxiliary indicators.[105] Under the new bonus system, if one index was not fulfilled, the bonus fund was reduced by 15 %. If both were unfulfilled, the fund was reduced by 30 %.[106] Tolerances for the degree of plan underfulfilment were not permitted for the enterprise. A third reason why enterprises still feared taut plans was that although bonuses for the enterprise workforce depended in the first instance upon the development of net profit, bonuses for management were based upon fulfilment of the two state plan indices, and upon their attention to contracts, exports, and the continuity of planned output.[107]

The weight attached to the two state plan indices provided the central authorities with a means of forcing enterprises to undertake technological change, particularly in priority, "structure-determining" areas. Yet this method could only function effectively if overall output planning was slack, and in the plan period 1969-1970 this does not seem to have been the case. Admittedly, it is not easy to evaluate how taut plans actually are. One method adopted by Keren with regard to the national economic plan was to take the past course of output growth as an indication of the possible: plans which demanded higher growth than in the past he branded as "taut". For each year from 1961 he compared the planned rate of growth to the actual rate of growth in the previous year of net material output, gross industrial output, consumption, and investment. His data showed the years 1969 and 1970 to be "taut".[108]

The GDR press also strongly indicates that the plan for 1970, given the weather conditions for that year, was taut. There were many reports of backlogs in plan fulfilment and of extra shifts and campaigns staged to help fulfil the half and full-year targets.[109] Mention was made in some articles of the "high targets" of the national economic plan, and in June at the Thirteenth Conference of the Central Committee, Günter Mittag criticised those who were saying the plan was "too high and must be slackened".[110] The journal Die Wirtschaft warned in July that in the struggle to make up plan backlogs it was not correct to concentrate solely on the fulfilment of the production plan and permit tasks in research, development and "transfer" to be neglected. Those directors who thought that "researchers and designers should help primarily in catching up on production backlogs" and conduct research and development on the side were multiplying the production difficulties of the future.[111] Paul Verner, secretary of the Central Committee, reported in December 1970 that "some of the high indices of the national economic plan" had not been reached. There were plan arrears particularly in sectors characterised by many-sided cooperation supplies. Serious disproportions had developed between suppliers and final producers and the backlog of contracts had increased from month to month. Exports were also behind schedule.[112]

Whereas regulations had been published in April and May of 1970 to continue the implementation of the bonus system introduced in 1969/70,[113] on 22 December 1970 a decree was published which in effect revoked this system.[114] Further details about the revision of the bonus system were published in February 1971.[115] Once again, enterprise success was measured relative to the plan, the operative period was one year, and the main index was planned net profit. Only if the plan for net profit and two auxiliary plans were fulfilled in 1971 could the enterprise draw on the full amount of the planned bonus fund.[116] The list of auxiliary indices differed from that of 1969/70 by the replacement of the two "structure-determining" indices and that for the "sale of finished products for the population" by that of "domestic contract fulfilment".[117] In 1971, as before, if one subsidiary target was not fulfilled, the bonus fund was reduced by 15 % and if both targets were not fulfilled, it was reduced by 30 %.

From 1972, the main index was planned commodity production. Planned net profit was an obligatory index. For every one percent over-, or underfulfilment of the main index, the bonus fund was changed by ± 1.5 %. The corresponding change for fulfilment of net profit was ± 0.5 %.[118] Additional payments to the bonus fund for the overfulfilment of the state plan index commodity production could only be made if two subsidiary indices were also fulfilled from the list: exports; marketed production for the population; production of important products; labour productivity.[119] These bonus rules did not provide much incentive for the enterprise to engage in technological change. Firstly, the main index "commodity production" oriented the enterprise towards output in a similar fashion to the main pre-1963 index. Secondly, the net profit indicator and fixed price system stimulated enterprise interest in manufacturing products with the highest profit margins, which were not usually new products; if the introduction of new technology could not be avoided, then enterprises were only interested in low costs once the price of a product had been set. In many enterprises of the machine building sector, management set cost-reduction targets for their design departments. Such targets were even prescribed for new products in their first year of manufacture and this induced the designer to "build reserves" into a new development so as to demonstrate "progress" in future cost reduction.[120] The tendency was undoubtedly strengthened in 1974 when enterprises had to undertake detailed cost planning and fulfil an index for prime cost reduction.

Thirdly, the supplementary indices offered little stimulus to technological change and enterprises did not need to fulfil them in order to receive at least the basic bonus. The indices for "mechanisation", "automation" and "labour productivity" still gave spurious results.[121] The most usual method of measuring labour productivity, for instance, was in terms of "commodity production" and manpower. With a constant denominator, labour productivity was highest when "commodity production" was highest, and this reinforced the interest of the enterprise in quantitative output and products with high prices. If the enterprise managed to economise on materials, the index either remained the same, or if a new product price was allotted, could even be lower than before.[122] With a constant numerator, labour productivity could be raised either by purchasing components from outside and thus "saving manpower", or by ignoring the time spent on technical standstills. A somewhat more satisfactory index of labour productivity was that based on individual enterprise net performance,[123] though even this did not adequately encourage economies of plant, material and energy, nor did it stimulate enterprises to be less interested in manufacturing products with high prices.[124]

The removal of science and technology from the list of auxiliary indices for the bonus fund meant that there was less motivation on the part of the enterprise to formulate slack plans for science and technology. This did not result, however, in the undertaking of projects promising significant technological change: instead interest centred on short-term R&D projects which would positively influence the two main indices for bonus payments. A number of critics therefore urged that the Plan for Science and Technology be given greater weight in the assessment of enterprise success.[125] A step in this direction was taken in 1975 by including for the first time in the list of state plan indicators such indices as the proportion of commodity production having the quality seals "Q" and "1", and the reduction of costs for waste, repairs and guarantee work.[126]

Interestingly a so-called "performance fund" was created in the enterprises in 1972 to reward workers for increases in labour productivity, cost savings as a result of reducing material and energy use, or improvement in quality.[127] This fund was problematic for it aimed at reducing "reserves" which were jealously guarded by the enterprises, and even if one of the three conditions for payment from the fund were fulfilled, the workforce might not receive much direct benefit: the fund could certainly be used to finance social projects, cultural activities for the employees, and subsidies for workers wishing to build their own homes, but it could also be used to implement rationalisation tasks in the enterprise and to support "joint measures with the local councils".

In the seventies, neither the bonus system nor the price system offered much stimulation to technological change. Criticisms were similar to those made under the system of detailed planning prior to 1963:

> As a result of measuring performance according to fulfilment of the commodity production plan and on this basis the calculated increase of labour productivity, the use of new technologies and the reduction of material consumption etc. often create problems for enterprise collectives. There are examples where enterprises have made significant contributions to increasing national economic effectiveness, but in the measurement of performance the impression is gained that the employees work worse after the introduction of new technology than before.[128]

One factor favouring technological change was that the national economic plans for 1971-1975 were slacker than those of 1969-1970 and it is unlikely that the reintroduction of the overbidding mechanism, in the form of "counterplans" altered this situation significantly.[129]

On the whole, the New Economic System was an imaginative policy concept and one promising greater stimulus to technological change than the detailed planning system. In practice, the financial instruments were imperfect and, as in the period prior to 1963, had to be complemented by pressure from the centre to promote major technological change. The reformers, however, had evidently hoped that the "economic levers" could be refined and would increasingly render such central pressures unnecessary. The financial incentive from the bonus and price systems was, in principle, highest between 1968 and 1970, yet in this same period the leadership attempted to bring about rapid technological change through a new "offensive strategy".

NOTES

1. "Richtlinie für das neue ökonomische System der Planung und Leitung der Volkswirtschaft", GBl, T II, Nr. 64 (1963), 453-498.

2. Ibid., 453.

3. For a general overview of this period see: G. Leptin and M. Melzer, Economic Reforms in East German Industry (London: Oxford University Press, 1978). In the interests of clarity the present chapter assumes for the most part an implicit three-tier model of industry, i.e. ministry-VVB-enterprise. It should be noted, however, that from 1968 combines were introduced which could either be directly subordinate to the ministry (i.e. at a level equivalent to the VVB) or subordinate to the VVB (i.e. at a level equivalent to the enterprise). The subject of combines will be treated more fully in Chapter Six.

4. Based on statistics assembled by: M. Keren, in "The New Economic System in the GDR: An Obituary", Soviet Studies, No. 4 (1972/73), 560.

5. G. Uhlemann, "Trotz reduzierter Vorgaben noch zu viele Planteile", Die Wirtschaft, Nr. 48 (1966), 15.

6. "Eine alte Kiste", Die Wirtschaft, Nr. 48 (1965), 17.

7. M. Böttcher, "Wo stehen wir mit der Planung im NÖS?", Die Wirtschaft, Nr. 48 (1965), 16.

8. Schmidt and Reinhold, "Probleme der Planung der wissenschaftlich-technischen Entwicklung", Die Wirtschaft, Nr. 2 (1966), 17.

9. See Keren, op. cit., note 4, 560 and 583.

10. G. Nitze and H.-J. Beckmann, "Der Plan Wissenschaft und Technik muß kontrollfähig sein", Die Wirtschaft, Nr. 20 (1974), 10.

11. Ibid.

12. "Wissenschaft und Technik brauchen langfristige Orientierung", Die Wirtschaft, Nr. 15 (1974), 2.

13. W. Schwarzbach, "Der Plan Wissenschaft und Technik muß stabil sein", Die Wirtschaft, Nr. 5 (1976), 21.

14. "Erkenntnisse aus der Erfüllung des Planes Wissenschaft und Technik", Die Wirtschaft, Nr. 21 (1974), 2.

15. "Forschungsplanung mit Risiko?", Die Wirtschaft, Nr. 18 (1974), 7. See also: K.U. Brossmann and A. Lange, "Forschung und Entwicklung mit Investitionen und Grundfondsreproduktion einheitlich planen", Die Technik, Nr. 10 (1975), 659; A. Lange, "Ökonomische Probleme der Überführung wissenschaftlich-technischer Ergebnisse in die Produktion", Einheit, Nr. 1 (1974), 104-108; A. Lange, D. Ivanov, R. Zimmermann and P. Wieczorek, "Intensivierung der Produktion und komplexere Planung des wissenschaftlich-technischen Fortschritts", Die Technik, Nr. 7 (1976), 450. NB: From 1976, science, technology and investments were to be planned on a five year basis.

16. A. Lange and D. Voigtberger, Überleitung von wissenschaftlich-technischen Ergebnissen (Berlin: Verlag Die Wirtschaft, 1975), 56 ff.

17. See Chapter Four.

18. Lange and Voigtberger, op. cit., note 16, 30-31; G. Klose, "Erfahrungen der Sowjetunion mit Koordinierungsplänen", Technische Gemeinschaft, Nr. 8 (1973), 21.

19. Lange and Voigtberger, op. cit., note 16, 30-31.

20. "Anordnung über die Finanzierung und Verrechnung der Forschungs- und Entwicklungsarbeiten in den Betrieben der volkseigenen Wirtschaft", GBl, T I, Nr. 82 (1957), 683 ff.

21. "Anordnung über die vorläufige Regelung zur Bildung und Verwendung des Fonds Technik in den dem Volkswirtschaftsrat unterstehenden Vereinigungen Volkseigener Betriebe für das Jahr 1964", GBl, T II, Nr. 89 (1963), 703-706. See also the later version: "Anordnung zur Bildung und Verwendung des Fonds Technik", GBl, T III, Nr. 26 (1965), 125 ff.; and H. Borchert, "Die Finanzierung des wissenschaftlich-technischen Fortschritts und die ökonomischen Hebel seiner schnellen Durchsetzung", Wissenschaftliche Zeitschrift der Martin-Luther-Universität, Halle-Wittenberg, Sonderheft 1964, 13 ff.

22. "Anordnung über die vorläufige Regelung...", op. cit., note 21, Section 2, Paragraph 4.

23. Ibid., 706.

24. H. Pöschel, "Forschung - Technik - Ideologie", Einheit, Nr. 3 (1966), 304; H. Kusicka, "Probleme der Planung, Leitung und materiellen Stimulierung von Forschung und Entwicklung", in A. Lange (ed.), Forschungsökonomie (Berlin: Verlag Die Wirtschaft, 1969), 151.

25. Die Wirtschaft, Beilage zur Ausgabe Nr. 28 (1965).

26. G. Feldmann and H. Brottke, "Praxisnahe Berechnung des

ökonomischen Nutzeffektes der Forschung und Entwicklung", *Die Wirtschaft*, Nr. 45 (1967), 10.

27. See C. Freeman, *The Economics of Industrial Innovation* (Harmondsworth: Penguin, 1974), Chapter 7; and K. Pavitt and W. Walker, "Government Policies Towards Industrial Innovation: A Review", *Research Policy*, *5* (1976), Section 5, 26-28.

28. H. Arnold, H. Borchert, A. Lange and J. Schmidt, *Die wissenschaftlich-technische Revolution in der Industrie der DDR* (Berlin: Verlag Die Wirtschaft, 1967), 145.

29. In a couple of early publications Pöschel cited the case of an enterprise in the electronics branch which paid lower wages and bonuses to a team responsible for the rapid development and transfer of a high quality product than it paid to other R&D collectives whose performance was only "average". The reason for this was that the first collective consisted of younger scientists and engineers. See: H. Pöschel and S. Wikarski, "Ideologische Probleme bei der Lösung der vom VI. Parteitag gestellten Aufgaben in Wissenschaft und Technik", *Einheit*, Nr. 6 (1963), 66; and H. Pöschel, *Leitung von Forschung und Technik und wissenschaftlicher Meinungsstreit* (Berlin: Dietz Verlag, 1964), 64. See further: E. Koschwitz, "Diskussionsbeitrag", in *Forschungsökonomie*, *op. cit.*, note 24, 240-241; E. Schade, "Maßstab der Prämierung muß die Planerfüllung sein", *Die Wirtschaft*, Nr. 33 (1967), 18.

30. Kusicka, *op. cit.*, note 24, 159.

31. L. Kannengießer, *Die Organisation der Beziehungen zwischen Wissenschaft und Produktion* (Berlin: Staatsverlag der Deutschen Demokratischen Republik, 1967), 85.

32. E. Koschwitz, "Prämierungsformen in Industrieforschungsinstituten", *Wirtschaftswissenschaft*, Nr. 7 (1968), 1110.

33. E. Dreyer, "Zur Anwendung des Prinzips der materiellen Interessiertheit in der Forschung und Entwicklung", *Die Wirtschaft*, Nr. 21 (1967), 9.

34. Schade, *op. cit.*, note 29, 18.

35. Koschwitz, *op. cit.*, note 32, 1120.

36. *Ibid.*, 1110, 1114.

37. "Anordnung über die auftragsgebundene Finanzierung wissenschaftlich-technischer Aufgaben und die Bildung und Verwendung des Fonds Wissenschaft und Technik vom 30. September 1968", *GBl*, T II, Nr. 110 (1968), 861; *Betriebsökonomie Industrie. Teil 2* (Berlin: Verlag Die Wirtschaft, 1971), 272 ff.

38. G. Prey, "Auftragsgebunden", *Deutsche Finanzwirtschaft*, Nr. 15 (1968), 1, 15.

39. *GBl*, T II, Nr. 110 (1968), 860.

40. *Ibid.*; and Prey, *op. cit.*, note 38, 15.

41. See e.g. R. Feller and S. Strauß, "Fonds Wissenschaft und Technik als Leitungsinstrument", *Die Wirtschaft*, Nr. 5 (1970), 5.

42. "Anordnung über die Bildung und Verwendung des Prämienfonds sowie des Kultur- und Sozialfonds in naturwissenschaftlich-technischen Forschungseinrichtungen der Deutschen Demokratischen Republik vom 14. Februar 1969", *GBl*, T II, Nr. 20 (1969), 142-143;

"Neue Prämienordnung", Deutsche Finanzwirtschaft, Nr. 10 (1969), F1.

43. "Richtlinien über die Preisbildung für wissenschaftlich-technische Leistungen vom 30. September 1968", GBl, T II, Nr. 110 (1968), 865-867.

44. GBl, T II, Nr. 20 (1969), 142.

45. L. Junker, "Finanzierung der Ingenieurbüros nach Effektivität", Sozialistische Finanzwirtschaft, Nr. 6 (1970), 51 f.

46. See for instance: H. Pöschel, "Ökonomie und Ideologie in Forschung und Technik", Einheit, Nr. 8 (1969), 933; R. Dittrich and J. Steiner, "Erfahrungen bei der Organisation des sozialistischen Wettbewerbes in produktionsvorbereitenden Bereichen", Sozialistische Arbeitswissenschaft, Nr. 5 (1973), 351; A. Lange and D. Voigtberger, Überleitung von wissenschaftlich-technischen Ergebnissen (Berlin: Verlag Die Wirtschaft, 1975), 86 and 57; K.-H. Behrendt and W. Heidel, "Leistungsbewertung und wissenschaftlich-technischer Fortschritt", Sozialistische Finanzwirtschaft, Nr. 11 (1973), 13.

47. R. Kühnemund and H.-J. Weihs, "Den wissenschaftlich-technischen Fortschritt auch mit Hilfe von Prämien beschleunigen", Arbeit und Arbeitsrecht, Nr. 16 (1976), 488.

48. W. Marschall, "Ökonomische Probleme des wissenschaftlich-technischen Fortschritts", Einheit, Nr. 4 (1973), 498; A. Lange, "Den Zyklus Wissenschaft-Technik-Produktion beherrschen (Teil I)", Die Technik, Nr. 1 (1973), 6.

49. Institut für Gesellschaftswissenschaften beim Zentralkomitee der SED (ed.), Wissenschaft und Produktion im Sozialismus (Berlin: Dietz Verlag, 1976), 265.

50. "Leistungsaufgaben und -instrumente für die Vervollkommnung der Planung der wissenschaftlich-technischen Arbeit", Die Wirtschaft, Beilage zur Ausgabe Nr. 31 (1973), 13; "Wissenschaft und Technik entscheiden wesentlich über das Wachstum der Volkswirtschaft", Statistische Praxis, Nr. 10 (1973), 498.

51. B. Schmiedeknecht, "Erfahrung mit der Verteidigungsanordnung", Die Wirtschaft, Nr. 46 (1974), 13.

52. "Erkenntnisse aus der Erfüllung des Planes Wissenschaft und Technik", Die Wirtschaft, Nr. 21 (1974), 2; A. Lange and W. Marschall, "Ökonomische Probleme bei der Überführung wissenschaftlich-technischer Ergebnisse in die Produktion", Wirtschaftswissenschaft, Nr. 2 (1975), 208; H. Sabisch and K. Hermsdorf, "Bedürfnis- und bedarfsgerechte Forschung und Entwicklung - ein Beitrag zur sozialistischen Intensivierung", Wissenschaftliche Zeitschrift der Technischen Universität Dresden, Nr. 1-2 (1976), 138.

53. "Verordnung über die Preisbildung nach der Güteklassifizierung des Deutschen Amtes für Meßwesen und Warenprüfung. - Preisbildungsverordnung, Güteklassifizierung - vom 29. Januar 1964", GBl, T II, Nr. 14 (1964), 117-118.

54. "Anordnung über die Gewährung von Gewinnzuschlägen und über die Beauflagung von Gewinnabschlägen vom 11. Februar 1964", GBl, T III, Nr. 15 (1964), 158-160.

55. S. Schauer, _Ökonomisch begründete Preisrelationen. Grundlagen und Methoden der Parameterpreisbildung_ (Berlin: Verlag Die Wirtschaft, 1965), 81; Schmidt and Reinhold, _op. cit._, note 8, 17.

56. _GBl_, T II, Nr. 14 (1964), 117.

57. "Anordnung über die Preisbildung für neu- und weiterentwickelte sowie für veraltete Erzeugnisse der metallverarbeitenden Betriebe", _GBl_, T II, Nr. 64 (1967), 423-428; "Anordnung über die Preisbildung für neu- und weiterentwickelte sowie veraltete Erzeugnisse der chemischen Industrie", _GBl_, T II, Nr. 122 (1968), 977-981; "Anordnung über die Preisbildung für Gußerzeugnisse, die nach neu- und weiterentwickelten sowie veralteten Fertigungsverfahren oder Gußwerkstoffen hergestellt werden", _GBl_, T II, Nr. 9 (1969), 83-87.

58. _GBl_, T II, Nr. 64 (1967), 426.

59. _GBl_, T II, Nr. 64 (1967), 428; _GBl_, T II, Nr. 122 (1968), 981; _GBl_, T II, Nr. 9 (1969), 87.

60. _GBl_, T II, Nr. 64 (1967), 426. NB: The following discussion is based mainly on the metalworking industry.

61. _Ibid._, 425-426. NB: In certain cases the enterprise itself could set the price.

62. F. Taut, "Neue Erzeugnisse - Inhalt des wissenschaftlich-technischen Fortschritts", _Die Wirtschaft_, Nr. 7 (1969), 7.

63. _GBl_, T II, Nr. 64 (1967), 427.

64. _Ibid._, 427, 428.

65. Taut, _op. cit._, note 62.

66. _GBl_, T II, Nr. 64 (1967), 424.

67. Jablonski, "Eine Verordnung ist noch kein Hebel", _Die Wirtschaft_, Nr. 29 (1970), 5.

68. E. Foth, "Produktion neuer Erzeugnisse genügend stimuliert?", _Die Wirtschaft_, Nr. 39 (1967), 14.

69. Taut, _op. cit._, note 62. G. Dinnies and R. Köhler, "Wirksamere Preisbestimmungen für neu- und weiterentwickelte sowie für veraltete Erzeugnisse der mvI", _Die Wirtschaft_, Nr. 1 (1971), 19.

70. _Ibid._

71. "Anordnung Nr. 3 über die Preisbildung für neue und weiterentwickelte sowie für veraltete Erzeugnisse der metallverarbeitenden Betriebe vom 28. Mai 1970", _GBl_, T II, Nr. 55 (1970), 417-418.

72. Jablonski, _op. cit._, note 67.

73. "Anordnung Nr. 3...", _op. cit._, note 71.

74. _GBl_, T II, Nr. 64 (1967), 428.

75. H. Höstermann, "Die Preisformen in der Industrie", _Deutsche Finanzwirtschaft_, Nr. 11 (1967), 10-11; "Anordnung Nr. Pr. 12 über die Preisformen bei Industriepreisen vom 14. November 1968", _GBl_, T II, Nr. 122 (1968), 971-973.

76. "Richtlinie zur Einführung des fondsbezogenen Industriepreises und der staatlichen normativen Regelung für die planmäßige Senkung von Industriepreisen in den Jahren 1969/1970 vom 26. Juni 1968", _GBl_, T II, Nr. 67 (1968), 497-504.

77. "Beschluß über Maßnahmen auf dem Gebiet der Leitung,

Planung und Entwicklung der Industriepreise vom 17. November 1971", *GBl*, T II, Nr. 77 (1971), 669-670.

78. H. Mann, "Die planmäßige Preisbildung als Instrument zur Förderung des wissenschaftlich-technischen Fortschritts", *Wirtschaftswissenschaft*, Nr. 6 (1975), 841; J. Berliner, *The Innovation Decision in Soviet Industry* (Cambridge, Mass.: The MIT Press, 1976), 293-295.

79. *GBl*, T II, Nr. 77 (1971), 669-670.

80. *Ibid.*, 671.

81. *Ibid.*, 673; "Anordnung über die zentrale staatliche Kalkulationsrichtlinie zur Bildung von Industriepreisen", *GBl*, T II, Nr. 67 (1972), 752-753.

82. *GBl*, T II, Nr. 77 (1971), 673.

83. T. Banse and H. Nick, "Gebrauchswert und Preisbildung", *Wirtschaftswissenschaft*, Nr. 6 (1975), 845.

84. N. Scholz, "Regulierung über den Preis?", *Die Wirtschaft*, Nr. 50 (1973), 11.

85. Mann, *op. cit.*, note 78, 840; Banse and Nick, *op. cit.*, note 83, 845.

86. H.J. Richter, "Parameterpreisbildung unterstützt wissenschaftlich-technischen Fortschritt", *Sozialistische Finanzwirtschaft*, Nr. 17 (1975), 16.

87. Banse and Nick, *op. cit.*, note 83, 846.

88. "Anordnung über die zentrale staatliche Kalkulationsrichtlinie zur Bildung von Industriepreisen", *GBl*, T I, Nr. 24 (1976), 321 ff.

89. "Richtlinie für das neue ökonomische System...", *op. cit.*, note 1, 472. NB: Leptin and Melzer, *op. cit.*, note 3, are contradictory on this point: see pages 20, 41, 48 and 121.

90. "Richtlinie...", *op. cit.*, note 1, 469.

91. *Lexikon der Industrie* (Berlin: Verlag Die Wirtschaft, 1970), 848.

92. Böttcher, *op. cit.*, note 7, 16-17. This article was published on 2 December 1965.

93. Keren, *op. cit.*, note 4, 565; and M. Keren, "The Rise and Fall of the New Economic System", in L. Legters (ed.), *The German Democratic Republic. A Developed Socialist Society* (Boulder, Co.: Westview Press, 1978), 68.

94. The law for the 1965 bonus fund was published on 14 April 1965. "Beschluß über die Grundsätze für die Bildung und Verwendung des einheitlichen Prämienfonds in der volkseigenen Wirtschaft im Jahre 1965", *GBl*, T II, Nr. 42 (1965), 297-298. The laws for the 1966 and 1967 bonus funds were published on 15 April 1966. "Beschluß zur Richtlinie für die Bildung und Verwendung des Prämienfonds in den volkseigenen und ihnen gleichgestellten Betrieben und den VVB der Industrie und des Bauwesens im Jahre 1967 sowie zur Übergangsregelung für das Jahr 1966 vom 7. April 1966", *GBl*, T II, Nr. 40 (1966), 249-256.

95. Keren, *op. cit.*, note 4, 578-579.

96. Tolerances for the degree of underfulfilment of the auxiliary indices were, however, permitted. "Beschluß über die Bildung und Verwendung des einheitlichen Prämienfonds in den

volkseigenen und ihnen gleichgestellten Betrieben der Industrie und des Bauwesens und in den VVB im Jahre 1964 vom 30. Januar 1964", _GBl_, T II, Nr. 10 (1964), 82.

97. _Ibid._, 82-86.

98. Böttcher, _op. cit._, note 7, 17.

99. _GBl_, T II, Nr. 10 (1964), 82.

100. F. Wesselburg, "Gewinn und leistungsgerechte Entlohnung des Leiters", _Wirtschaftswissenschaft_, Nr. 10 (1965), 1589.

101. "Verordnung über die Bildung und Verwendung des Prämienfonds in den volkseigenen und ihnen gleichgestellten Betrieben, volkseigenen Kombinaten, den VVB (Zentrale) und Einrichtungen für die Jahre 1969 und 1970 vom 26. Juni 1968", _GBl_, T II, Nr. 67 (1968), 492; R. Wiesner, "Prämienfonds unterstützt Strukturpolitik", _Deutsche Finanzwirtschaft_, Nr. 21 (1968), F1.

102. Based on: "Beschluß über die Grundsatzregelung für komplexe Maßnahmen zur weiteren Gestaltung des ökonomischen Systems des Sozialismus in der Planung und Wirtschaftsführung für die Jahre 1969 und 1970 vom 26. Juni 1968", _GBl_, T II, Nr. 66 (1968), 449-450.

103. _GBl_, T II, Nr. 67 (1968), 490; _GBl_, T II, Nr. 66 (1968), 449-450.

104. _GBl_, T II, Nr. 67 (1968), 491.

105. R. Wiesner, "Zu einigen Hauptaufgaben der Prämierung", _Die Wirtschaft_, Nr. 39 (1968), 15.

106. _GBl_, T II, Nr. 67 (1968), 491. NB: The wording of the law on this point is not very clear. See Wiesner, _op. cit._, note 105, for a better exposition.

107. _GBl_, T II, Nr. 67 (1968), 492.

108. Keren, _op. cit._, note 4, 577.

109. For the first half year see: "Auch bei Konsumgütern Planerfüllung sichern", _Die Wirtschaft_, Nr. 15 (9. April 1970), 2; "Verstärkt den Kampf um die Planerfüllung organisieren", _Die Wirtschaft_, Nr. 21 (14. Mai 1970), 2; "Kampf gegen Planrückstände und Schaffung des wissenschaftlich-technischen Verlaufs gehören zusammen", _Die Wirtschaft_, Nr. 23 (4. Juni 1970), 3; "Kontinuierlich die Planaufgaben erfüllen", _Die Wirtschaft_, Nr. 24 (11. Juni 1970), 2; "Die Plandisziplin konsequent einhalten und festigen", _Die Wirtschaft_, Nr. 28 (9. Juli 1970), 2.

110. "Die Durchführung des Volkswirtschaftsplanes im Jahre 1970", _Neues Deutschland_ (11. Juni 1970), 3.

111. "Überleitungsprozeß mit höchster Effektivität gestalten", _Die Wirtschaft_, Nr. 29 (16. Juli 1970), 2.

112. P. Verner, "Aus dem Bericht des Politbüros an die 14. Tagung des ZK der SED", _Neues Deutschland_ (10. Dezember 1970), 4-5. See also: "Die Entwicklung der Volkswirtschaft im Jahre 1970, dem letzten Jahre des Perspektivplans 1966 bis 1970", _Neues Deutschland_ (22. Januar 1971), 3.

113. "Grundsatzregelung für die Gestaltung des ökonomischen Systems des Sozialismus in der Deutschen Demokratischen Republik im Zeitraum 1971-1975", _Die Wirtschaft_, Nr. 18 (29. April 1970), Beilage 14, 17-20; "Rechtsvorschriften zur Durchführung der in der Grundsatzregelung für die Gestaltung des ökonomischen Systems

des Sozialismus in der Deutschen Demokratischen Republik im Zeitraum 1971 bis 1975 enthaltenen Aufgaben", Die Wirtschaft, Nr. 19-20 (7. Mai 1970), Beilage 15, 85-89.

114. "Beschluß über die Durchführung des ökonomischen Systems des Sozialismus im Jahre 1971 vom 1. Dezember 1970", GBl, T II, Nr. 100 (1970), 731 ff.

115. "Verordnung über die Planung, Bildung und Verwendung des Prämienfonds und des Kultur- und Sozialfonds für das Jahr 1971", GBl, T II, Nr. 16 (11. Februar 1971), 105.

116. Ibid., 106. See also Napierkowski, "Die Planung des Prämienfonds als absoluter Betrag", Die Wirtschaft, Nr. 13 (1971), 14.

117. GBl, T II, Nr. 16 (1971), 106.

118. "Verordnung über die Planung, Bildung und Verwendung des Prämienfonds und des Kultur- und Sozialfonds für volkseigene Betriebe im Jahre 1972", GBl, T II, Nr. 5 (1972), 49.

119. Ibid.

120. E. Garbe, "Brauchen wir 'Änderungs-Reserven'?", Die Wirtschaft, Nr. 49 (1972), 7. See also the comments by Mann, Banse and Nick in the section on prices.

121. See the comments on these indices in Chapter Four.

122. See e.g. H. Nick, Intensivierung und wissenschaftlich-technischer Fortschritt (Berlin: Dietz Verlag, 1974), 63-65.

123. W. Müller, "Wie die Arbeitsproduktivität messen?", Die Wirtschaft, Nr. 14 (1972), 6.

124. R. Heuer and H. Marx, "Zur Messung der Arbeitsproduktivität", Die Arbeit, Nr. 4 (1973), 34.

125. See e.g. M. Heinelt, G. Klose and J. Zobel, "Plan und Leistungsbewertung müssen den technischen Fortschritt besser stimulieren", Die Wirtschaft, Nr. 18 (1974), 8; S. Hornich, "Die Pläne Wissenschaft und Technik zum wichtigsten Bestandteil der Volkswirtschaftspläne machen", Die Wirtschaft, Nr. 29 (1974), 4-5; W. Leupold, "Rationeller Einsatz der Mittel und Kräfte in Wissenschaft und Technik", Energietechnik, Nr. 4 (1973), 149.

126. "Höhere Qualitätsziele im Plan 1975", Die Wirtschaft, Nr. 34 (1974), 2.

127. "Anordnung über die Planung, Bildung und Verwendung des Leistungsfonds der volkseigenen Betriebe", GBl, T II, Nr. 42 (1972), 467-468; "Anordnung Nr. 2 über die Planung, Bildung und Verwendung des Leistungsfonds der volkseigenen Betriebe", GBl, T I, Nr. 7 (1974), 66.

128. Hornich, op. cit., note 125, 4. See also Nick, op. cit., note 122, 35-36.

129. H. Koch and F. Paschke, "Erfahrungen aus der Arbeit mit Gegenplänen gehören in die Rahmenrichtlinie", Die Wirtschaft, Nr. 18 (1974), 9.

6
The Offensive Strategy for Technological Change, 1967/68–1971

The "offensive strategy" was an ambitious attempt by the centre to accelerate technological change and appears to have been borne of impatience with the instruments of the New Economic System; an impatience which first surfaced through comments of leading GDR politicians during 1967. Delivering the closing address to the Seventh Party Congress in April of that year, Ulbricht began by noting that the GDR had made significant progress towards becoming a highly effective, modern economy since the previous congress in 1963, and he attributed this mainly to the NES. Later in his speech, however, he complained that party policy on the development of the economy had been implemented unsatisfactorily and too slowly.[1] He urged that the future structure of the economy be characterised by products and processes which permitted efficient manufacture at the highest scientific-technical level, assured a high rate of profitability on exports, raised productivity and technological levels in important branches, and facilitated the efficient use of domestic raw materials and energy sources.[2]

At a seminar for "leading cadre of the party, state and economy" in September 1967, Ulbricht said that plan proposals for the following year were insufficiently orientated towards top-quality world market products: too much research was still being conducted on tasks that had already been solved in the world.[3] In November 1967 Günter Mittag complained that the Plan for Science and Technology was not fulfilled in many enterprises and institutes; that R&D periods were too long; that the results transferred into production were often of a mediocre standard; and that in the "defence" of projects, users and others who might contribute to a real "scientific conflict of opinion" and a comparison with the "world level" were often deliberately excluded. R&D establishments frequently had no clear orientation towards the needs of the national economy, works' directors tended to neglect R&D for short-term financial advantages, and the GDR suffered considerable economic loss from the long lead times. An example of the latter, he said, was the process "Solvent Polymerisation of Acrylonitrile": although a research team of the Institute for Research on Fibrous Materials had presented results on this process as early as 1961, large-scale use of the process had been postponed several times and was now planned for 1970 in the "VEB Synthetic Fibre Works 'Friedrich Engels' Premnitz"; in the meantime, however, an English concern had brought a further developed, more productive process onto the market.[4] At the Ninth Conference of the Central Committee in 1968, Ulbricht noted that a joint conference on computers had been held in West Berlin between the Technical University of West Berlin and MIT of the USA. A terminal in West Berlin had been connected with the computer centre in Boston, over 6,000 km away, and a demonstration given of automatic literature research.

> I would be interested to know what the responsible comrades in the government think about this. We, at any rate, think that it is not just a coincidence or something of no importance, but a challenge. Is it not about time to hold an analogous conference at the Humboldt University or the University for Economics showing, in closest cooperation with the most advanced institutes of the Soviet Union (the Institutes in Moscow, Novosibirsk, Kiev), what the most modern electronic data processing machines can do in research and technology?[5]

The ambitious aim of the GDR, conceived in 1967 and launched in 1968, was to leapfrog the current top levels of science and technology in the world. Political slogans changed accordingly: no longer was the GDR to "catch up and overtake" but to "overtake without catching up". The "offensive strategy for technological change" had two main components: priorities for technological change and concentration in industry. The use of priorities greatly strengthened the role of central planning in certain branches and enterprises at the expense of the infant New Economic System. "Pioneer achievements" were called for in growth or "structure-determining areas". Forecasts were to sketch the likely development of the economy and technology, and help locate areas for "pioneer achievement" which were then to be promoted by detailed central planning. This whole process was to be facilitated by a more efficient structure of research and production achieved through the concentration of industry.

PRIORITIES FOR TECHNOLOGICAL CHANGE

The areas designated as priority were reportedly derived from a newly formulated "structure-political conception" which outlined the future developmental objectives of the economy.[6] Research areas, products and technologies expected to be important for competition on the world market or for raising the technological level of the economy were to receive special promotion. Unfortunately no detailed data have been published on the "structure-determining" tasks, products and resource allocation owing to the GDR's obsession for secrecy.[7] As Günter Mittag explained in October 1968, the "strictest state discipline" necessary for preserving secrets also applied to implementing the structural policy. The "class enemy" was trying to find out what developments were envisaged in the structural policy, what new results had been obtained in research and development, where important new industrial plants and automation projects were located, and how the GDR's foreign trade could be penetrated.[8]

Nevertheless, the growth branches may be discerned from production statistics. Table 6.1 contains the average annual growth in branch production for the years 1969 and 1970, and shows that several branches had growths considerably higher than the overall industrial average of 6.3 %. In the chemical industry, chemical fibres had the highest growth (13.8 %), followed by plastics (12.7 %), mineral oil, gas and coal processing (8.7 %) and pharmaceuticals (7.9 %). In the machine and vehicle building industry, the branch with the highest growth was chemical plants (19.6 %), followed by agricultural machines (15 %), pneumatic and refrigeration machines (13.4 %), metal constructions (10.9 %) and machine tools (10.4 %). In the electrotechnical, electronics, and instrument building industry, output of data processing and office machines increased by 18.9 %, that of electronics, fine mechanical and optical products by 14 %, and that of measurement and process control technology by 13.1 %.

The priority "structure-determining" branches were, it seems, mainly ones in which the GDR was especially weak: it will be recalled from Chapter Two that in 1968 the GDR appeared to have a substantial technological lag with the FRG in most of these branches. Chapter Three also showed that the GDR

Table 6.1
Average Annual Growth of Gross Production in 1969-1970 (%)

Energy and Fuel Industry	3.2
Chemical Industry	7.3
Mineral Oil, Gas and Coal Processing	8.7
Inorganic and Organic Basic Chemicals	1.4
Pharmaceuticals	7.9
Plastics	12.7
Rubber and Asbestos	6.6
Chemical Fibres	13.8
Industrial Chemical and Special Products	5.0
Metallurgy	7.4
Building Materials	7.6
Electrotechnical, Electronics, Instrument Building	11.0
Electrotechnical	7.2
Electronics	14.0
Measurement and Process Control Technology	13.1
Data Processing, Office Machines	18.9
Fine Mechanical and Optical	14.0
Machine and Vehicle Building	6.5
Chemical Plants	19.6
Building-, Building Materials-, and Ceramic Machines	8.4
Pneumatic and Refrigeration Machines	13.4
Machine Tools	10.4
Polygraphic Machines	7.5
Textile, Clothing, Leather Machines	7.7
Rail Vehicles	5.8
Road Vehicles, Tractors	5.4
Ships	8.4
Agricultural Machines	15.0
Conveying and Lifting Equipment	8.2
Motors, Pumps, Compressors	8.3
Machine Parts	6.5
Metal Constructions	10.9
Foundries and Forges	0.0
Metal Goods	5.4
Light Industry	6.7
Textile Industry	3.8
Food Industry	4.7
Total Industry	6.3

Source: Calculated from data in: DDR Wirtschaft. Eine Bestandsaufnahme (Frankfurt am Main: Fischer Verlag, 1974), 355-357.

in 1964 had a low R&D effort relative to the FRG in chemicals, plastics, data processing, and in electrical machine and apparatus construction. An energetic ideological campaign was waged enjoining research establishments and enterprises to bring forth "pioneer and supreme achievements" particularly in the priority branches, and censuring in public the complacent. The party leadership's conception of "pioneer achievements", however, was rather unrealistic. According to Walter Ulbricht, these were to surpass by a considerable margin the current top level of science and technology in the world and were to "open up fully new, previously unknown paths" superior to those already in existence and thus provide "far-reaching opportunities for high rates of increase of societal labour productivity".[9]

> Therefore we speak deliberately not of technical progress but of the scientific-technical revolution. This is to express unequivocally that it is not a matter of gradual, step-by-step change of existing technologies. Instead what we need are completely new technological processes ...[10]

The leadership appeared to assume that radical new discoveries and inventions could be made rapidly, especially by setting higher standards or trying harder, and overlooked the high degree of uncertainty involved in radical technological change, the essential role of creativity, the economic significance of incremental advance, and the length of time needed for major innovations to go from conceptualisation to commercial introduction.

The "structure-political conception" was purportedly based on forecasts. These and other "rational" techniques such as operations research, computer methods, and scientific management had received increasing attention from the party leadership from the early sixties.[11] At the Seventh Party Congress in April 1967 Ulbricht called for a considerable expansion and improvement of forecasting activity.[12] In October of that year he complained that the State Planning Commission, the ministries and the VVBs had not yet solved the "main problem" of specifying processes, product groups and technologies to be promoted in order to make them the basis of a highly productive structure in the economy.[13] To look into this problem, special forecasting groups were to be set up in the State Planning Commission and other economic agencies. A source published in 1968 states that the Council of Ministers had sixteen forecasting groups involving over 2,000 academics, state and economic functionaries.[14] The groups apparently worked quickly, for in summer 1968 the "structure-political conception" was completed and the "nomenclature" of priority products and technologies worked out.

> The path to success is only via thorough forecasting work; the recognition in good time of developmental tendencies in science and technology; continual and relentless comparison with the world level; rapid transfer of scientific results into production; and the rationalisation and automation of complex production processes.
>
> Every time the breakthrough to the top world level has been attained, the same experience has been made: the precondition for supreme achievements is scientific management activity assisted by data processing and operations research.[15]

No doubt the party's enthusiasm for these techniques was influenced by the extensive interest both of the Soviet Union[16] and the industrial capitalist countries in such methods. Probably the techniques were viewed as an important source of Western, and especially West German, success in technological change, and since science could only develop unfettered in a socialist society, regarded as being potentially all the more effective in the GDR.

There was a widespread idea in the GDR that forecasting would greatly reduce the uncertainty about the future and combined with other "scientific" methods would permit accurate long-term planning. According to Ulbricht:

> In the fulfilment of our assignments we proceed from the preparation of forecasts. This makes possible a correct structural and science policy, and allows the exact stipulation of societal tasks.[17]

Yet previous experience with long-term planning had not been entirely satisfactory. The first five-year economic plan for 1951-1955 had to be changed four times; the second five-year plan for 1956-1960 only became law in December 1957 and was discontinued in 1959; the perspective plan for 1959-1965 was discarded in 1962/63; and the economic perspective plan for 1964-1970 only became law in 1967 after the completion of the third price reform. Rather than concluding from this experience that a flexible, incremental planning approach was necessary, for there was an essential uncertainty about the future, the official aim according to Bartsch and Kraft was to use scientific methods to "open up the future". "It must be quite clear: supreme achievements result from highly qualified forecasts..."[18] It was surely a sign of the times that these authors commended "systematic heuristics" as a source of inspiration.[19] Another writer referred to cybernetics, operations research, electronic data processing and network techniques as a "dialectical unity".[20] Günter Mittag stated in 1970:

> Forecasting work is to be organised in such a way that in the course of the work itself new, previously unknown ideas are gained on scientific-technical and technological possibilities. It is a case of obtaining bold solutions. Forecasting work is therefore genuine scientific preliminary research.[21]

Closer attention of the party leadership to the practice, as opposed to the advocacy, of forecasting in the West would have revealed their expectations to be unfounded. In general, forecasts by government and industry were not particularly accurate: the greater the degree of uncertainty about the future, the less reliable the forecasts. Nor could forecasts provide a quicker route to scientific-technical knowledge than basic and applied research itself. Empirical surveys indicated that few Western companies actually engaged in systematic forecasting activity, especially as it was extremely difficult to identify all areas of change important to an organisation and to assess the potential impact of these changes. Amongst companies that did undertake formal forecasts, there was little evidence of effective integration with planning or decision making.[22]

The hopes pinned by the party leadership on forecasting and other "scientific" methods were to be disappointed.[23] Forecasts might be divided broadly into two related categories: those of scientific-technical opportunities and those of market opportunities or customer needs. With regard to the first, it was not possible to specify what opportunities would be available fifteen years hence and extrapolate these backwards to construct economic plans. Most GDR forecasts of scientific-technical development in fact projected current technologies onto the future, and this orientation towards "the world level of today" was criticised for giving insufficient impetus to the "attainment of "pioneer achievements".[24] With regard to forecasts of markets and needs, it was frequently the case that the customer himself did not know what he wanted and the problem was compounded when management personnel did not have very high professional abilities. The technical director of an enterprise specialising in automation equipment pointed out that he could not expect management in branches such as the food industry and agriculture to state their needs for automation equipment to 1980 as these personnel seldom had qualifications higher than that of the secondary technical school.[25] Forecasts of market possibilities were also impaired by the generally low level of market research; by the difficulty of procuring comparative data on world market prices and on the operating costs of competitor firms; and by the slowness in pricing new GDR products.

There were a number of other problems affecting both categories of forecasts. Complaints were made about insufficient personnel and fragmentation of forecasting activity. Persons delegated to work out forecasts in the enterprise sometimes had to do this on top of their normal duties. Forecasts were regarded by some in industry as "frivolities" and were "knocked together" within a few weeks simply because something had to be produced.[26] Not surprisingly the information which was passed on to the forecasting groups was sometimes criticised as being so vague as to be useless.[27] Difficulties also arose from the fact that a forecast was often the resultant of a number of component forecasts. If the latter were not produced on time, or were of insufficient quality, the overall forecast would be of little use. A further obstacle was that the forecasting period usually adopted spanned from 1968 to 1980: although a standard period was important for the coordination of the various forecasts, it had the grave disadvantage that factors specific to the particular product were often ignored.[28] In view of the above problems, it is likely that the economists and functionaries responsible for formulating the "structure-political conception" and the "nomenclature" of priority technologies based their ideas less on any creative "highly qualified forecasts" than on careful surveys of developments in countries at the "world level" of technology, especially in the FRG.

Priority areas were placed under a regime of detailed central planning and this had several advantages. Firstly, it helped to counteract enterprise apathy or resistance to the introduction of new or improved technology. An important instrument in this respect was mentioned in the previous chapter: structure-determining tasks could be set as state plan indices, the non-fulfilment of which resulted in heavy reductions of the bonus fund for the enterprise workforce and possibly no bonuses for management. Secondly, enterprises entrusted with structure-determining tasks were given priority in balances, orders and contracts. Thirdly, the planners were not encumbered with details on the workings of the whole economy as they had been up to 1963: non-priority areas remained within the framework of the New Economic System.

Disadvantages were that the enterprise would strive for "slack", and the central authorities for "taut", state plan indices. If too many tasks were labelled "priority", the non-priority areas would be neglected and the planners would lose their overview. Furthermore, it is difficult to avoid the suspicion that the planning of structure-determining tasks, whilst suitable for promoting technological diffusion and imitation, was too protracted and bureaucratic to permit the rapid and flexible action inherent in any "offensive" technological change. To support this argument it is instructive to consider the mechanism of "structure-determining" planning in a little more detail.

The first stage of the planning involved the State Planning Commission suggesting a "nomenclature" of priority products and technologies to the Council of Ministers. After confirmation, the Council of Ministers passed on projects to the appropriate minister who was equipped with special powers of directive and was responsible to the Council for the overall planning, balancing and realisation of the projects. He selected certain VVBs, combines and enterprises to perform the tasks and appointed "task managers". The task managers were responsible for the planning and execution of the tasks at the lower level of the economy, and in particular had to ensure optimal supply and cooperation relations. They were furnished by the minister with "clearly defined, task-based and temporally limited powers of authority" which could transcend their usual spheres of competence.[29] The enterprises and combines had to prepare so-called "structure-concrete plan data" which were then sent up to their immediate superior agency for coordination. The agency would then confer with the enterprises and prepare a reference which was sent, together with the "structure-concrete data", for "defence" to the minister responsible. Prior to this "defence" procedure, the minister was obliged to conduct "coordinating" discussions with the directors of the various agencies involved. If these discussions also necessitated decisions by the Council of Ministers, he had to prepare the decisions by submitting "economic alternatives" after consultation with the

State Planning Commission, central and regional management. Once the "structure-concrete data" had been successfully defended before the minister, he provisionally confirmed an outline for the planning and execution of structure-determining tasks. He also passed on recommendations to the State Planning Commission for inclusion of the tasks in the state plan. The procedure, however, was not yet complete. The next step involved the minister "defending" his proposal for structure-determining tasks before the Council of Ministers. Taking part in the proceedings were the management of other central state agencies and the chairmen of regional councils involved in the tasks. Finally, on the basis of the defended proposals and the advice of the State Planning Commission, the Council of Ministers stipulated plan-tasks for the enterprises and economic agencies to extend over the next few years.[30]

CONCENTRATION IN INDUSTRY

"Fragmentation" of research and production had long been an issue for criticism in the GDR, and up to 1967 a certain amount of industrial concentration had taken place, both via enterprise fusion and the formation of VVBs. From 1968, however, an intensive drive was made to concentrate production and research. Three types of enterprise concentration were promoted: the fusion of enterprises into larger ones; the formation of combines; and an increased division of labour and specialisation through cooperation relations between enterprises. Integral to these types was the creation of socialist "large-scale research".

Table 6.2 contains a formal comparison of industrial concentration in the two Germanies from 1963 to 1975, the measure being the proportion of the country's employed persons working in enterprises having more than 1,000 employees.[31] Several points may be noted. Firstly, even before the increased concentration drive beginning in 1968, industrial production in the GDR appears to have been more concentrated than in the FRG. Secondly, whereas the level of concentration in the GDR continually increased over the period 1963-1975, in the FRG it remained roughly constant. And thirdly, dividing the development in the GDR into four-year periods, the greatest increase in the level of concentration occurred between 1968 and 1971: 8.9 %. Between 1964 and 1967, there was an increase of 2.6 %, and between 1972 and 1975 a slight increase of 0.6 %. Although this comparison must remain a tentative one, particularly in view of

Table 6.2
Percentage of Total Employees[a] in Large Industrial Enterprises[b] in GDR and FRG

	GDR	FRG
1963	51.3	38.8
1964	52.6	39.1
1965	53.8	39.1
1966	54.6	38.6
1967	55.2	37.6
1968	56.2	38.2
1969	60.6	39.4
1970	63.3	40.2
1971	65.1	39.4
1972	66.9	38.4
1973	65.9	39.0
1974	66.5	39.3
1975	67.5	38.6

Sources: Statistical Yearbooks of the GDR and FRG (1965-1977).

Notes: a) GDR figures exclude apprentices.
b) Enterprises having more than 1,000 employees.

possible differences in definitions,[32] it does serve to illustrate the policy pursued in the GDR of actively promoting the concentration of production as compared with the rather static situation in the FRG.

A more detailed idea of the extent of industrial concentration in the GDR over the period 1968-1971 may be obtained from Table 6.3. This shows the proportion of employees and number of enterprises by size of enterprise. Over the period, the proportion of employees in all enterprise groups fell except in those having more than 1,000 employees. Similarly, the number of enterprises in all enterprise groups having fewer than 1,000 employees fell. All enterprise groups larger than 1,000 employees increased both their proportion of employees and the number of enterprises. The biggest increases, however, were in the enterprise group having more than 5,000 employees: between 1968 and 1971 the proportion of employees increased from 20 % to 27 % and the number of enterprises by 20.

Great importance was accorded to the establishment of combines. Prior to 1968 there were only a few combines in the GDR; by the end of 1972 there were 120, and these accounted for 60 % of the commodity

Table 6.3
Industrial Concentration in GDR between 1968 and 1971

Enterprise Size[a]	Prop. of Employees[b] 1968	Prop. of Employees[b] 1971	Number of Enterprises 1968	Number of Enterprises 1971
Up to 25	2.2	2.1	4,156	3,864
26 - 50	3.7	3.3	2,853	2,559
51 - 100	5.3	4.5	2,117	1,812
101 - 200	7.7	5.7	1,502	1,147
201 - 500	12.6	9.3	1,126	845
501 - 1,000	12.4	10.0	499	405
1,001 - 2,500	21.6	22.5	401	406
2,501 - 5,000	14.5	15.6	121	133
Over 5,000	20.1	27.0	62	82
	100	100	12,837	11,253

Sources: St.J.DDR 1970, 110; St.J.DDR 1973, 122.

Notes: a) Measured in terms of the number of employees.
b) Excluding apprentices.

production of the centrally managed sector of industry.[33] Thirty-seven of the combines were directly subordinate to a ministry rather than a VVB. Lange in 1970 attempted to compile a list of possible benefits from combine formation. In general he saw economies resulting from the centralisation of various functions such as R&D, patent work, technology policy, purchasing, storage, marketing and management. The concentration and specialisation of R&D personnel should, he thought, lead to a more rapid reaction to customer and market needs, to a shortening of development time, and to a more rapid transfer of results into production. In production the increased division of labour would lead to a higher piece number and permit the changeover to batch and flow process technology.[34] According to the enterprise journal Die Wirtschaft in 1968:

> The implementation of the Economic System of Socialism as a whole, the mastery of the scientific-technical revolution, and not least the class struggle with imperialism requires that higher organisational forms be sought and the most appropriate of these applied.
> For many economic units, the combine is the most sensible higher organisational form. With

> its technical-economic and cadre potential it is in a much better position than the individual enterprise to master all phases of the reproduction process, and at a time of increasing confrontation with the world market, to direct its efforts towards the manufacture of products capable of competing on this market.
> Combine formation and therefore the concentration of production leads to higher profitability and productivity.[35]

In a later article, the journal claimed similarly that combines made it possible to manufacture "complex product systems with the greatest effectiveness and to attain the highest scientific-technical level".[36] Günter Mittag pronounced that combines achieved "the unity of science and production".[37]

Combines were to be equipped with large research centres, and in general both industrial and academic research were to be reorganised into "socialist large-scale research", characterised by large research establishments, large teams and a reduced number of projects. The relevance of research work to industry was to be encouraged by research contracts. The scale of finance from the state budget was to be substantially reduced, and R&D establishments had to charge their customers for scientific-technical results. Günter Prey, the Minister for Science and Technology, put the case for "large-scale research" and its relationship with industry as follows:

> To attain the requisite new quality of planning, management and organisation of scientific-technical work, we need a socialist organisation of science corresponding to the top world level; a new type and style of scientific-technical research and production. This means the development of socialist large-scale research and its close, harmonious integration with socialist large-scale production.[38]

One GDR commentator thought that only large-scale research would enable the GDR "to master the scientific-technical revolution",[39] and Die Wirtschaft even stated that the concentration of research was "the most important precondition for raising scientific creativity".[40]

According to a book written by the economists Graichen and Rouscik, the GDR had 1,800 R&D establishments in 1968. 83 % of these employed "fewer than 100 persons; 53 %, fewer than 10 persons; and many, only one to two persons".[41] The authors stated that it was nowadays clear that one person could have only a limited overview of the "world level" for fairly simple products and a small group could only develop imitations: it took a large R&D collective to bring out a revolutionary new development. The optimal size of an R&D establishment was viewed internationally as being between 200 and 300 employees, with 100 persons constituting a "critical mass". After 1975 the optimum would lie between 400 and 500 persons, whereby the "critical mass" depended heavily on the economic importance and character of the research work. In the future, research centres employing several thousand persons would come into existence in the GDR concerned with areas of importance to the economy.[42]

But were the assumptions that concentration led to better conditions for technological change and to higher economic effectiveness supported by empirical evidence in GDR publications? Concerning the first question no serious investigation appears to have been undertaken prior to the concentration drive launched in 1968. At most, writers pointed to the concentration of R&D in large Western firms:

> The concentration of research and development in the developed capitalist industrial countries is further advanced than that of industrial production. Both within and between the monopoly groups it is a matter of achieving the best utilisation conditions for invested capital. In the USA, 90 % of all research expenditures are concentrated in 500 concerns. The research expenditure of the largest English chemical concern, Imperial Chemical Industries (ICI), exceeds that of all the universities of the country taken together. Five large firms in the Netherlands account for half of the total research expenditure of the country.[43]

The increasing concentration of R&D in large West German firms over the years may be seen from Table A.11 on the number of R&D personnel in firms of three size-groups between 1964 and 1971. Whereas in 1964, 5.1 % of total R&D personnel was employed in firms having fewer than 500 employees, by 1971 this proportion had reduced to 2.5 %. On the other hand, firms having 2,000 and more employees employed 85.3 % of total R&D manpower in 1964, and by 1971 this proportion was 88.5 %. The same trends may be seen for the development of "scientific personnel".

The statistical evidence on the concentration of R&D in large firms in the advanced capitalist countries is indisputable, yet it does not mean that small firms cannot play a useful role in accomplishing technological change. This view was taken by a number of Western economists who published investigations during the early sixties on size of firm and inventive output. Jewkes, Sawers and Stillerman pointed out that most industrial laboratories in the United States were small - more than half employed fewer than fifteen people - and argued a case for the significant role of the individual inventor. They found that more than half of their sample of sixty-one "important" twentieth century inventions resulted from individuals who either worked on their own behalf with little in the way of resources and assistance, or were employed by institutions such as universities where they were free to pursue their own ideas without hindrance.[44] Hamberg also found that of twenty-seven important inventions from the decade between 1946 and 1955, only seven had originated in large industrial laboratories; the remainder came from independent inventors, small firms, a lone agronomist, and three universities. In his view, large industrial laboratories were more likely to produce improvement inventions than radically new inventions.[45] Scherer found on ranking US firms by size of sales that inventive inputs (R&D employment) and outputs (patents) increased less than proportionally with sales. His data suggested that smallness was not necessarily an impediment to the creation of patentable inventions.[46]

There is a difference, of course, between the creation of inventions and the development and marketing of these inventions, i.e. the completion of innovation. About half of Jewkes' "private" inventions were developed and brought to the market by large corporations. An empirical study on the contribution to innovation of various sized firms in the United Kingdom was completed by Freeman in 1971. He found that although post-war innovation (1945-1970) had been dominated by large firms, the growth of professional R&D had not eliminated the contribution of small firms to industrial innovation. Specifically, between 1945 and 1970, "giant" firms (more than 10,000 employees) accounted for 54 % of the innovations in his sample; "large" firms (more than 1,000 employees) for 79 % of the innovations; "medium" firms (200-999 employees) for 11 %, and "small" firms (1-199 employees) for 10 %.[47]

Important in the question of whether a small firm can play a useful role in industrial innovation or not is the branch of industry. In some branches, only large firms are in a position to command the facilities, the finance and manpower necessary to undertake expensive or large-scale innovation. In other branches, a small firm might have the advantage of strong motivation, flexibility, and effective coupling of the various stages of the innovation process. The notion that enterprise concentration in the GDR would lead to better conditions for technological change was probably something of an overgeneralisation. The GDR leadership was right in recognising the important role to be played by large enterprises in technological change, though perhaps mistaken in appearing to assume that small enterprises could not play an effective role.

With regard to concentration and higher economic effectiveness, three articles were published in the GDR during 1966 and 1967, i.e. prior to the intense concentration drive, on the "optimal size of the enterprise".[48] All three pointed out that the question of optimal enterprise size had received little attention in socialist economic literature. Kuciak stated that the development of enterprise size was "mainly determined" by two economic "laws": the "law" of mass production and the "law" of plant scale. These economies of scale meant that given identical production programmes, larger enterprises were economically more efficient than small enterprises. His criterion for the "optimal" enterprise size was the best "world level" in labour productivity and in manufacturing costs; an optimally sized enterprise permitted the GDR to sell her products on the world market with profit. Kuciak, however, did not provide statistical data to support his assertions, nor were factors such as size of market, demand, location of enterprise, labour force situation, and availability of materials integral to his notion of "optimal" enterprise size; rather, these were external factors which would "relativise" the optimum. From this point of view, he argued that for certain mass manufactures, the optimal enterprise size would be so large that a small country like the GDR would not be able to utilise the capacity to the full:

> If we manufacture such products in smaller enterprises, then we cannot achieve the world level in labour productivity and costs. The higher the proportion of such non-optimal enterprises in the national economy, the greater the economic loss, for example in foreign trade.[49]

He saw two solutions to this. Either to do without such products and specialise in areas where the GDR could achieve the "world level" in costs, or to secure sufficient demand through international division of labour and specialisation.

Böttcher cited some conclusions of an empirical investigation conducted in certain VVBs. One "interesting discovery" was that a group of enterprises of less than medium size in one VVB worked considerably more effectively in terms of profit and other indicators than all other enterprise size-groups, including the four largest enterprises. Data from four VVBs of different industrial sectors showed that there was no fixed "optimal" enterprise size. The most favourable costs in the four sectors were held by enterprise groups of different sizes. This emphasised the importance of the branch of industry and type of product. In Böttcher's view, every enterprise size was optimal when it had the most appropriate production structure. By "production structure" he apparently meant the type and quantity of production, and the technological level and specialisation of the production process. If the size of the enterprise was fixed, it would be optimal once the appropriate production structure had been found; if the production structure was fixed, the size of the enterprise should be adjusted so as to show the lowest costs.[50]

Kuczmera also held that there was no fixed optimal size for an enterprise. In his view, basic criteria for optimal enterprise size were capacity, sales potential and operating costs or profit. In addition there were a number of other factors to be borne in mind such as the location of the enterprise, availability of materials, labour force and features specific to the industrial branch.

> As a result of the multiplicity of factors to be considered, a purely mathematical derivation of optimal enterprise size is not possible. The optimal enterprise size is not a fixed quantity.[51]

He criticised the inclination of some designers towards "overdimensional plants and aggregates". These required considerable investment funds, and often after a long construction period no longer corresponded to the "highest world level" of technology. He also underlined the importance of a distribution of enterprise sizes in the economy:

> The sole construction of large enterprises would be inappropriate with regard to solving the tasks of the socialist economy. Only economically suitable proportions between large, medium and small enterprises will lead rapidly to an efficient industry.[52]

In short, the few GDR publications on optimal enterprise size which appeared before the concentration wave beginning in 1968 were by no means unanimous that concentration led to higher economic effectiveness. Nor was this assumption supported by empirical evidence. Both in the question of appropriate enterprise size for optimal economic effectiveness and for best conditions for technological change, the party leadership seemed to adhere to an implicit philosophy of "bigness".

This philosophy also lay behind the creation of combines. They had a number of advantages over the VVB-enterprise system but it is difficult to subscribe to the unconditional acclamation of them contained in most GDR publications. Each combine was run on the principle of "single management": one director had overall responsibility for the whole of its operations. This, together with the fact that a combine could be a vertical integration of "supply" and "customer" enterprises, meant that the director was potentially in a more advantageous position to coordinate production than the management of the horizontal VVB had been. He was also in a better position to strengthen the link between science and production: he was nearer production than VVB or ministerial administrations and was more likely to know about the possibility of modernising plants or of introducing new products; he had a research centre at his disposal; and he could avail himself of certain managerial devices to avoid discontinuities in the process of technological change. "Transfer teams", for example, composed of designers, technologists and production workers accompanied the project from the testing stage to the beginning of production.[53] "Task management", as noted above, could also be applied for important projects approved by the Council of Ministers. The rights and duties of the task manager were formally stipulated in a law published in March 1970; and these included powers of directive against the various organisations involved in the "task" and the authority to form special-purpose work groups and "socialist collectives".[54]

Combines may have relieved the overall volume of planning and paper work. Although "structure-determining tasks" involved a substantial amount of planning and decision making, it is possible that certain competences previously exercised by the ministry or VVB were devolved to the combine director, and that plan coordination between enterprises of different branches became less problematic. The latter would certainly have been the case in the "vertical" combine because previously cooperation between enterprises of different branches had necessitated extensive paper work as plans had run over different VVBs and ministries. A further advantage of combines was that they could be readily oriented towards the aims and priorities of the centre. Significant in this respect was that by 1972 virtually one-third of the GDR's combines were directly subordinate to the ministry rather than the VVB. The central agencies could thus intervene more easily to promote technological change and especially "structure-determining tasks"; whether this possibility was used wisely depended, of course, upon the level of information and professional competence of the central personnel involved.

Against these advantages must be set a number of practical problems. Firstly, combines were often geographically fragmented and this hindered both intimate knowledge of the component enterprises, and the transfer of manpower from one enterprise to another.[55] Secondly, the position of the combine director was somewhat anomalous: he was simultaneously a representative of the ministry or VVB and something of an entrepreneur; he was supposed to execute central aims and exercise individual initiative. Given that he had his contracts and state plan indices to fulfil, his enthusiasm for promoting significant technological change not specifically provided for via state indices would not have been high; and in having to strike a balance for the work of the combine research institute between the extremes of "hobby research" and production work, he might have been tempted towards the latter. Not least, it is quite likely that old conflicts of interest between enterprise and VVB or ministry reappeared between enterprise and combine administration. The advantage of combines most frequently stressed in GDR publications was the greater efficiency of large-scale operations. But whether large-scale production actually resulted in economies of scale depended on the branch of industry, type of product, size of market, location of enterprises etc., and whether large-scale research was appropriate depended heavily on the sort of research involved.

If the leadership's implicit philosophy of "bigness" was not supported by empirical evidence, what foundation did it have? One factor was no doubt policy tendencies in the Soviet Union. In a speech to the Central Committee of the Soviet Communist Party in 1971, Brezhnev stated:

> The increased concentration of production is becoming a necessity. Experience shows that only large associations are in a position to concentrate the requisite number of specialists,

> to assure rapid technical progress, and to make better and more complete use of all resources. The course aimed at creating associations and combines must be implemented resolutely.[56]

And in 1969 Die Wirtschaft wrote:

> As the experience of the Soviet Union confirms, supreme achievements in research and development require the concentrated application of personnel and resources onto the main tasks of the state structural policy. Consequently it is necessary to overcome the fragmentation still found in research and development and to change over from individual research to socialist large-scale research. Decisive preconditions for this concentration in science and technology and the rapid transfer of results into production are being created by the formation of efficient socialist combines.[57]

Another factor, important both in the Soviet Union and the GDR, was Marxist-Leninist theory. Marx spoke in Das Kapital of economies resulting from concentrating the means of production. "In a large factory with one or two central motors, the costs of these motors do not increase in the same proportion as their horsepower...". Further savings were made on items such as buildings, fuel, and lighting whereas "other conditions of production remain the same, whether used by few or by many.[58] Other factors being equal, labour productivity depended on the scale of production.[59] Lenin stated that with the development of capitalism, "many fragmented production processes fuse to a single societal production process"[60], and in his "Draft Plan for Scientific-Technical Work" he advocated "from the viewpoint of the most modern large-scale industry and especially trusts, a rational amalgamation and concentration of production in a few large enterprises".[61]

> Socialism has been created through large-scale mechanical industry. And if the working masses introducing socialism do not understand that they should adapt their institutions to the functioning of large-scale mechanical industry, then one cannot even speak of an introduction of socialism.[62]

A third and related factor was how analysts and policy makers in Eastern Europe perceived developments in the West. They were probably just as impressed by Western proponents of large-scale firms as they were by Western advocates of forecasting and other instruments of "scientific management". Galbraith's view that only large firms were capable of producing and promoting modern technology undoubtedly met with widespread consensus, despite the anathema accorded to his notions on the "convergence of the systems". A fourth factor concerns the psychology of Walter Ulbricht. During the period 1967-1971 he was at the zenith of his career and power. His seventy-fifth birthday on 30 June 1968 was occasioned by elaborate festivities and he was lauded by the SED as being "the most significant German politician of the century".[63] Given this background, it is conceivable that Ulbricht simply associated bigness with greatness, and regarded cost-benefit calculations as irrelevant if not insulting.

The leadership's policy of concentrating research and production did not go unquestioned. Certain scholars, either hinted or warned against forming large research groups. Dietrich Wahl, participant in a group at the Academy of Sciences researching into the organisation of research, wrote in 1969 that experience showed the most effective research collectives to have between seven and thirteen scientists, and up to twenty-five persons altogether. This, he said, did not argue against "socialist large-scale research", rather it indicated that in large-scale research centres the most favourably sized groups had to be formed.[64] G. Block warned in the same year that figures on the "optimal" size or "critical mass" of a research institution should not be overrated: there were no hard and fast rules and it was important to proceed from the concrete situation of the particular industrial branch or concern.[65] Some people felt that "large-scale research" could only be afforded by a large country, and that the GDR did not have the resources to make "pioneer achievements".[66] Ulbricht charged, however, that to

> contest the possibility and necessity of large-scale research was nothing less than to doubt the prospective existence of socialist large-scale industry, and to accept a doleful future in which we content ourselves with an average industry and a mediocre labour productivity.[67]

Yet doubt about socialist large-scale industry is what a number of GDR economists had. In 1968 two GDR economists published papers which in effect questioned the official policy of concentrating production. Rouscik stated that analyses in the GDR, in socialist and capitalist industrialised countries showed that for a number of branches, large enterprises did not achieve the best economic results.[68] In many branches of the manufacturing industry, there was no direct connection between greater plant efficiency and enterprise size.[69]

> The optimal enterprise size can be exhibited by large, medium as well as small enterprises. This depends on the industrial branch, its products and product range, and the level of development of technique and technology. The large enterprise alone is no criterion for the optimum ...[70]

New fields of activity, he said, were continually opening up for optimal small and medium-sized enterprises in a dynamically growing economy. Enterprise fusion involved many problems and the main thing was to be clear about the necessity of such a measure and not to indulge in "fusion at any price".[71] Kuczmera produced figures on labour productivity over the various branches with respect to enterprise size. According to these figures, the most productive enterprise size-groups were not always the largest; the situation varied according to the particular industrial branch. Furthermore, the majority of enterprises were not concentrated in the most productive enterprise size-groups:

> For example in metallurgy the highest labour productivity is in enterprises having up to 25 employees. There are only 3 of this size in the whole of metallurgy, whereas 10 enterprises have more than 5,000 employees - an enterprise size which specifically in metallurgy proves to be least productive.[72]

The papers by Rouscik and Kuczmera, published in a supplement to the journal Die Wirtschaft, apparently prompted a number of people in industry to question the government's policy of concentration. Several weeks later the same journal published an editorial stating that the papers had provoked "contradiction in questions of principle" and:

> Some theses must be put right because they would obstruct a concentration of production. Indeed several economic practitioners who are not clear about the necessity of a rapid concentration of production in the GDR for the mastery of the scientific-technical revolution have cited the contents of the supplement, and in addition misinterpreted several other things.[73]

To help clarify matters, a couple of other authors were asked to give their opinions on the question of concentration and optimal enterprise size. The first article published was written by Professor Dr. Karl Hartmann of the Party University "Karl Marx".

Hartmann criticised Rouscik and Kuczmera for not showing the extent and rate at which the formation of "larger more effective structural units" was necessary and stated their articles evoked "illusionary ideas" about the level of concentration attained in the GDR compared with what was necessary. Firstly, he said, the use of science as a productive force required the concentration of resources: forecasts and international experience had shown that "supreme achievements" were only possible on this basis, and "Comrade" Walter Ulbricht had emphasised the importance of concentrating research potential onto structure-determining areas within the shortest possible time. Secondly, the "scientific-technical revolution" was increasingly characterised by automation and a serious obstacle to automation was the existing "fragmentation" in industry. Thirdly, Hartmann pointed to the formation of "many large enterprises in the Soviet Union, particularly in the structure-determining sectors", and stated that the concentration and centralisation of production and capital in the USA, Japan and West Germany was more extensive than ever.

> From this point of view also, we cannot set the tempo arbitrarily. The class struggle with West German imperialism intensifies the necessity of raising the level of concentration, specialisation and cooperation in our production.[74]

He put forward two reasons why the statistics did not necessarily show the best economic results for the largest enterprises: one was that the advantages of large-scale production would only become apparent when the "Economic System of Socialism" was more fully developed; another was that it was important to use modern methods of enterprise management. In his opinion, however, it was not sufficient to view the concentration process from the standpoint of the individual enterprise. Important was the operation and concentration of all the enterprises within a particular production system. Rouscik and Kuczmera had not taken this perspective, and had failed to discuss the relationship between enterprise cooperation and concentration, which had been dealt with at the Third Plenum of the Central Committee of the Socialist Unity Party. For as Walter Ulbricht had said: "The way to the implementation of concentration and specialisation in socialist society leads via socialist cooperation."[75]

Hartmann's article was followed two weeks later by one from Kuciak. He explained that the contributions published in the Die Wirtschaft supplement contained statements which one could not agree with, as they gave a "false orientation". He thus wished to elaborate to a certain extent upon Professor Hartmann's "extensive critique". His first point was that although optimal enterprise size was a component aspect of the concentration process, it was not identical to it: important, as Ulbricht had emphasised, was the role of cooperation. Another point was that statistical material on labour productivity by enterprise size was only of limited use, not least since enterprises even in the same industrial branch often had quite different production structures, and so too much weight could not be attached to Kuczmera's figures. For Kuciak, as for Hartmann, however, the necessity of concentration appears to have been mainly a matter of political faith:

> Marx proved that the development of the division of labour and the concentration of production is a condition and consequence of the development of the production forces, and thus assumes the character of a law.[76]

With one eye on the "classics", another on the Soviet Union, an occasional glance towards the West, and symptoms of megalomania, the party leadership ordered an intensive concentration drive in industry. Admittedly, the decrees on concentrating industry stipulated that before the decision was taken to form a combine or fuse enterprises, an "exact" calculation of the economic effect had to be worked out.[77] This was a sensible provision but in practice it appears to have been eclipsed by propaganda on the necessity of concentration, and the higher-level economic functionaries "concentrated" enterprises irrespective of whether greater economic effectiveness would be obtained. Behrendt, a researcher in the Ministry of Finance, called in 1970 for reliable and detailed proof of the economic effectiveness of combine formation before taking the decision to go ahead.[78]

The concentration of R&D is reflected in the following figures: in 1966 about 80 % of all R&D personnel were employed in establishments having up to 50 persons; in 1975 almost two-thirds worked in establishments of more than 200 employees.[79] However there was much uncertainty about the function of "large-scale research centres" and about the nature of "large-scale research" generally. Publications and comments by the party leadership often gave the impression that the whole of research was to be restructured along "large-scale" lines, if not by creating large-scale research centres then by introducing cooperation relations between various establishments. Understandably, this provoked anxiety in the scientific-technical community for it broached the prospect of having to relinquish a particular research area, join a team, and possibly move to a different town.[80] There was also a general apprehension about becoming merely small cogs in a great machine, with little recognition of individual performance.[81] The economist Alfred Lange criticised those who thought that "large-scale research" was to be organised "everywhere" and would be completed with the formation of research collectives of between thirty and fifty people: in his view this type of research should be restricted to a small number of research complexes.[82]

A rather vague article was published on the "Strategy of Large-Scale Research Centres" by Die Wirtschaft. These centres, it said, were responsible for the rapid attainment of "pioneer and supreme achievements", should develop into the main performers of research in industry, and ought to avoid taking on R&D work connected with current production from the enterprises. The article, however, placed the onus on the director of the centre to "develop appropriate methods on how to manage the research process", pointing out that one would be just as mistaken to cling to current forms of economic organisation as one would to orientate a large-scale research centre solely towards the further development of what already exists.[83]

Ulbricht criticised in 1970 the confusion about the nature of large-scale research and its centres, although he did little to clarify matters. Referring to the chemical industry, he said that in some cases massive investment programmes had been drawn up for large-scale research centres which involved the construction of new buildings, offices and laboratories but which merely "conserved the economic forms and methods of research of the past".[84] The Minister for Science and Technology, Günter Prey, complained about the way industrial ministers and combine directors met their responsibilities for "questions of science policy". In one case, a large-scale research centre had even been delegated the task of conducting all the planning work on science and technology for the VVB. According to Prey there were some "strange" and indeed "extreme" ideas about the role of the large-scale research centre in the combine. At one extreme was the notion that such centres should take over everything connected with research, design and project preparation. At the other extreme was the redesignation of even modest research capacities as "large-scale research centres": in practice they were working with "name plates" and "proclamations".[85]

THE STRATEGY AS A WHOLE

The party leadership had apparently identified or was aware of the areas of technology in which the GDR was weak, and was concerned to improve the situation. Propaganda about "pioneer achievements" was possibly a useful pressure towards new technology; forecasting activity helped to render assumptions and alternatives explicit; priority planning served to counterbalance enterprise resistance to technological change; and the creation of combines made it possible to strengthen the link between research and production, and to reduce the volume of paper work. Nevertheless, the leadership had unrealistic ideas about achieving radically new technologies, about forecasting, about the effectiveness of large-scale operations, and in general about the psychology and sociology of invention and innovation.

Given the high growth target for industry,[86] the strategy probably involved too many structure-determining tasks and too much reorganisation. The overall output plan was not slack enough to permit its fulfilment at the same time as the implementation of radical technological change and the creation of combines, larger enterprises and cooperation groups. The concentration and specialisation of production led to delays in starting up production and in supplies. The priority promotion of "structure-determining areas" resulted in a neglect of non-priority areas and products, which in turn also caused supply difficulties for the priority branches. The whole situation was exacerbated by the bad weather conditions of 1969 and 1970. Bottlenecks appeared in the energy, building, transport and agriculture branches and the ambitious strategy resulted in 1970 in a new growth crisis: the economy was in a state of disequilibrium.

At the inception of the offensive strategy, some had warned that such a policy might lead to disproportions in the economy, although Lorenz and Haker of the State Planning Commission had brushed these warnings aside with the point that legal guidelines had been set which would guarantee a proportional development of the economy.[87] Whereas the aim of the strategy had been to overtake the West in certain areas of technology, the increasing problems of supply, energy, transport etc. meant that the enterprise manager had to devote most of his attention to the fulfilment of his production plan and had little interest in pushing through technological change.[88] At most, the strategy promoted technological diffusion and imitation, although scientific-technical tasks not designated priority, such as improving product quality, were almost certainly neglected. In December 1969, for instance, wholesalers had to return 17 % of all black and white television sets to the manufacturer. In February and March 1970, the proportion of defect colour television sets per delivery lay between 7 % and 30 %. Statistics from seven sales outlets revealed that of 1,800 washing machines sold in the first quarter of 1970, every second one was faulty.[89]

In his report to the Fourteenth Conference of the Central Committee in December 1970, Paul Verner stated that it was necessary to make certain "corrections" to the current policy. These were not corrections of the economic system itself: he did not share the mistaken view of some comrades, including economic functionaries, who felt that the Economic System of Socialism (the later appellation for the NES) had not proven itself; rather they were corrections of "certain exaggerated notions and wishes". He did not spell these "corrections" out in detail although he made clear that the structural policy was to remain, care being taken to assure a "proportional development of the national economy", and that besides the development of large-scale research, the "daily work of researchers and innovators in the enterprises and combines" was to receive stronger support.[90]

In 1971, however, more drastic changes were made: both the "offensive strategy" and the New Economic System were dropped and there was a change of top political leadership: the heir apparent, Honecker, took over from Ulbricht the post of party secretary. Whilst the reasons for Ulbricht's replacement must remain a matter for speculation, important factors were probably his economic policy from 1968, his intransigence over Moscow's efforts to improve relations with Bonn, and his immoderate ambition for the GDR both within Comecon and as a "model" of socialism. The new leader's main task now was to restore stability and improve the lot of the consumer.

As noted in Chapter Five, there was a return to the central planning methods used prior to 1963: the number of plan indices for the enterprises was raised and indirect management through economic "levers" limited; the responsibilities of the enterprises and VVBs were reduced and those of the top-level economic agencies increased. The concentration of production as may be seen from Table 6.2 came to a virtual standstill. So too did the concentration of research. The concept of "large-scale research" was dropped and centres were renamed "research centres". Official propaganda now emphasised "intensified" use of existing technologies rather than great leaps forward in new technology. The amount of attention paid in publications to forecasts and other "scientific" methods of management decreased and the leading role of the party was reasserted. No longer was there much brave talk of "mastering the scientific-technical revolution"; the more modest aim was now to accelerate "scientific-technical progress".

If the "New Economic System" was not dead by the end of 1971, as some Western observers have asserted, it had at least been considerably dismantled.[91] The NES was only allowed a run of just over four years before the leadership launched its new, overambitious, "offensive" strategy. In the formulation of this strategy, perhaps too much faith was placed in the efficacy of "rational" methods, refined central planning, policy tendencies in the Soviet Union, and the teachings of Marxist-Leninist theory. The implementation of the strategy under conditions of taut planning confounded the development of what the reformers had thought would be a more efficient and less bureaucratic method of managing the economy.

> The successful development of the GDR is mainly due to the fact that the Socialist Unity Party has creatively applied the Marxian teachings of socioeconomic societal formation, the laws and principles of socialist society which were further developed by V.I. Lenin into a generally valid scientific theory of socialist construction.[92]

NOTES

1. W. Ulbricht, Die gesellschaftliche Entwicklung in der Deutschen Demokratischen Republik bis zur Vollendung des Sozialismus. VII. Parteitag der Sozialistischen Einheitspartei Deutschlands, Berlin, 17. bis 22. April 1967 (Berlin: Dietz Verlag, 1970), 14, 86, 124-125.

2. Ibid., 104-105.

3. W. Ulbricht, "Der Weg zur Durchführung der Beschlüsse des VII. Parteitages der SED auf dem Gebiet der Wirtschaft, Wissenschaft und Technik", Die Wirtschaft, Nr. 41 (1967), 4.

4. G. Mittag, "Von der 3. Tagung des Zentralkomitees. Aus

dem Bericht des Politbüros an das 3. Plenum", Neues Deutschland (24. November 1967), 5-6.

5. W. Ulbricht, "Die weitere Gestaltung des gesellschaftlichen Systems des Sozialismus", Die Wirtschaft, Beilage zur Ausgabe Nr. 44 (31. Oktober 1968), 17.

6. A. Rüger, "Die Bedeutung 'strukturbestimmender Aufgaben' für die Wirtschaftsplanung und -organisation der DDR", Deutsches Institut für Wirtschaftsforschung, Sonderhefte, Nr. 85 (1969), 12; see also "Beschluß über die Grundsatzregelung für komplexe Maßnahmen zur weiteren Gestaltung des ökonomischen Systems des Sozialismus in der Planung und Wirtschaftsführung für die Jahre 1969 und 1970", GBl, T II, Nr. 66 (1968), 435 ff; and "Beschluß des Staatsrates der Deutschen Demokratischen Republik über weitere Maßnahmen zur Gestaltung des ökonomischen Systems des Sozialismus", Die Wirtschaft, Beilage zur Ausgabe Nr. 18 (1968), 4 ff.

7. For some fragmentary information see: A. Rüger, op. cit., note 6, 13; G. Mittag, "Unsere sozialistische Planwirtschaft ermöglicht höhere Effektivität", Die Wirtschaft, Beilage zur Ausgabe Nr. 18 (1968), 7; and Ulbricht, op. cit., note 1, 104-105.

8. G. Mittag, "Meisterung der Ökonomie ist für uns Klassenkampf", Neues Deutschland (27. Oktober 1968), 5.

9. W. Ulbricht, "'Überholen ohne einzuholen' - ein wichtiger Grundsatz unserer Wissenschaftspolitik", Die Wirtschaft, Nr. 9 (1970), 8-9.

10. Ibid., 8.

11. For two early criticisms about the neglect of such methods see: M.A. Ardenne, Wege zur Steigerung der Weltmarktfähigkeit unserer industriellen Erzeugnisse (Berlin: Verlag Die Wirtschaft, 1963), 25; L. Rouscik, "Das neue ökonomische System der Planung und Leitung der Volkswirtschaft fördert die Wissenschaft und Technik", Die Technik, 18, Nr. 12 (1963), 766.

12. Ulbricht, op. cit., note 1, 101.

13. Ulbricht, op. cit., note 3, 5.

14. "Strukturpolitik", Die Wirtschaft, Nr. 27 (1968), 2.

15. "Im größeren Umfang Pionier- und Spitzenleistungen erzielen", Die Wirtschaft, Nr. 37 (1969), 2.

16. See for instance: W. Gluschkow and G. Dobrow, "Die wissenschaftliche Prognose", Die Wirtschaft, Nr. 3 (1969), 30; B. Saizew and B. Lapin, "Prognostizierung von Wissenschaft und Technik", Die Wirtschaft, Nr. 28 (1969), 23.

17. Ulbricht, op. cit., note 3, 3.

18. H. Bartsch and G. Kraft, "Prognostik ist kein Hilfsprozeß", Die Wirtschaft, Nr. 47 (1970), 13. These authors worked in the Central Committee's "Institute for Social Sciences".

19. Ibid., 14.

20. M. Bunge, "Kybernetik - Operationsforschung - Netzwerktechnik - EDV", Die Wirtschaft, Nr. 33 (1968), 19.

21. Quoted in H. Lilie, "Aus den Erfahrungen der bisherigen Prognosearbeit", Die Wirtschaft, Nr. 21 (1970), 5.

22. See e.g. J.P. Dory and R.J. Lord, "Does TF really work", Harvard Business Review (November-December 1970), 16-28, 168; E. Mansfield, "Technological Forecasting", in T.S. Khachaturov

(ed.), Methods of Long-Term Planning and Forecasting (London: Macmillan, 1976), 334-349; K. Pavitt, "Analytical Techniques in Government Science Policy", Futures (March 1972), 5-12; C. Freeman, C. Cooper and K. Pavitt, "Policies for Technical Change", in C. Freeman and M. Jahoda (eds.), World Futures (London: Martin Robertson, 1978), 207 ff.

23. For some criticism by Ulbricht in 1969 on forecasting and operations research see: "Aus den Materialien der 10. Tagung des ZK der SED", Die Wirtschaft, Beilage zur Ausgabe Nr. 20 (15. Mai 1969), 3.

24. For two criticisms of this sort, see: S. Tannhäuser, "Konzentration des Forschungspotentials", Die Wirtschaft, Beilage zur Ausgabe Nr. 18 (1968), 16; and "Aus dem Bericht über die Ergebnisse der Untersuchung zu Problemen der Weiterentwicklung der Planung in den Betrieben und Kombinaten im ökonomischen System des Sozialismus", Die Wirtschaft, Beilage zur Ausgabe Nr. 18 (1968), 22.

25. "Antworten aus der Praxis", Die Wirtschaft, Nr. 33 (1968), 8-9.

26. "Nur ein modernes sozialistisches Planungssystem sichert Pionier- und Spitzenleistungen", Die Wirtschaft, Nr. 3 (1970), 3.

27. E. Hasler, "Praktische Probleme der Prognosearbeit", Die Wirtschaft, Nr. 33 (1968), 8.

28. Ibid., 8-9.

29. H.-J. Lorenz and R. Haker, "Die Planung, Bilanzierung und Realisierung volkswirtschaftlich-strukturbestimmender Aufgaben", Die Wirtschaft, Nr. 30 (1968), 5.

30. See for example: "Beschluß...", op. cit., note 6, 436 ff.; "Kommentare zum Entwurf der Grundsatzregelung für komplexe Maßnahmen zur weiteren Gestaltung des ökonomischen Systems des Sozialismus in der Planung und Wirtschaftsführung für die Jahre 1969 und 1970", Die Wirtschaft, Beilage zur Ausgabe Nr. 19 (1968); Lorenz and Haker, op. cit., note 29, 5-6; E. Melms, "Praktische Fragen der vorrangigen Planung, Bilanzierung und Realisierung volkswirtschaftlich strukturbestimmender Aufgaben", Die Wirtschaft, Nr. 37 (1968), 7.

31. Lack of data prohibited the comparison of enterprises having say more than 5,000 or 10,000 employees.

32. It must remain rather tentative for several reasons. Firstly, the figures for the GDR do not reflect the concentration resulting from the formation of cooperation groups. Secondly, there are problems connected with the meaning of the term "enterprise": it is not clear from the definitions of this term in the statistical yearbooks of the two Germanies whether the two concepts are equivalent. The GDR definition mentions that an enterprise may consist of several geographically separate enterprise components. If these "components" were in West Germany, they might be considered as separate enterprises, for according to the FRG definition, an enterprise is a "geographically separate branch of a firm". (See definitions in the Statistical Yearbooks of the GDR and FRG.) Thus the GDR figures may be inflated relative to those of the FRG owing to a possible difference in definition.

Another factor tending to inflate the GDR figures is the formation of combines after 1968. West German statistics would classify a "combine" as a "firm" consisting of several "enterprises". GDR statistical yearbooks for 1968 and later years are rather unclear as to whether a "combine" is regarded as one enterprise or several in tables on enterprise size. In the former case, Table 6.2 would reflect the concentration of production as a result of enterprise fusion and combine formation. In the latter case, it would reflect only enterprise fusion. The yearbooks for 1968, 1969 and 1972-1975 state that in tables on "the number of enterprises and their development" a combine is considered as a group of enterprises. Nevertheless, the yearbooks for 1970 and 1971 state in addition that in the tables on enterprise size, a combine is regarded as one enterprise unit (St.J.DDR 1972, 110 and St.J.DDR 1973, 110). Comparing the figures for 1970 and 1971 with those for the other years after 1968, it would be reasonable to assume that the same criterion for tables on enterprise size was used in all years after 1968.

33. D. Graichen and B. Siegert, Sozialistische Wirtschafts-, Wissenschafts- und Leitungsorganisation (Berlin: Verlag Die Wirtschaft, 1974), 53.

34. H. Lange, "Die Konzentrationseffekte und ihre Realisierung, Grundlagen der höheren Effektivität der Kombinate", Wissenschaftliche Zeitschrift der Hochschule für Ökonomie, Berlin, Nr. 2 (1970), 111-118.

35. "Konzentrationsprozeß", Die Wirtschaft, Nr. 32 (1968), 2.

36. "Kombinatsbildung und Effektivität", Die Wirtschaft, Nr. 24 (1969), 11.

37. See S. Wikarski, "Wesen und Aufgabe der sozialistischen Großforschung", Einheit, Nr. 3 (1969), 381.

38. G. Prey, "Einige Führungsprobleme der sozialistischen Großforschung", Effekt, Nr. 3 (1969), 5.

39. H.-U. Gramsch, "Durchsetzung sozialistischer Großforschung", Die Wirtschaft, Nr. 14 (1969), 3.

40. "Wissenschaft und Technik vor neuen Aufgaben", Die Wirtschaft, Nr. 5 (1969), 9.

41. D. Graichen and L. Rouscik, Zur sozialistischen Wirtschaftsorganisation (Berlin: Verlag Die Wirtschaft, 1971), 225.

42. Ibid.

43. H. Kusicka and W. Leupold, Industrieforschung und Ökonomie (Berlin: Dietz Verlag, 1966), 34.

44. J. Jewkes, D. Sawers and R. Stillerman, The Sources of Invention (London: Macmillan, first edition 1959, second edition 1969), 117 and 73 in second edition.

45. D. Hamberg, "Invention in the Industrial Research Laboratory", Journal of Political Economy (April 1963), 95-115.

46. F.M. Scherer, "Firm Size, Market Structure, Opportunity, and the Output of Potential Inventions", American Economic Review, 55 (1965), esp. 1103-1105; see also F.M. Scherer, "Unternehmensgröße und technischer Fortschritt", Die Aussprache, Nr. 7 (1969), 169-174.

47. C. Freeman, The Economics of Industrial Innovation

(Harmondsworth: Penguin Books, 1974), 210-218; also C. Freeman, Innovation and Size of Firm, Occasional Papers No. 1 (Science Policy Research Centre, Griffith University, March 1978).

48. M. Böttcher, "Die optimale Betriebsgröße - Scheinproblem oder Mittel ökonomischer Diagnose?", Die Wirtschaft, Nr. 48 (1966), 10-12; G. Kuciak, "Ökonomische Gesetze und Betriebsgröße", Die Wirtschaft, Nr. 7 (1967), 11-12; M. Kuczmera, "Die optimale Betriebsgröße - eine Notwendigkeit für unsere Volkswirtschaft", Die Wirtschaft, Nr. 7 (1967), 12.

49. Kuciak, op. cit., note 48.

50. Böttcher, op. cit., note 48.

51. Kuczmera, op. cit., note 48.

52. Ibid.

53. H. Nick, Intensivierung und wissenschaftlich-technischer Fortschritt (Berlin: Dietz Verlag, 1974), 168-169; A. Lange and D. Voigtberger, Überleitung von wissenschaftlich-technischen Ergebnissen (Berlin: Verlag Die Wirtschaft, 1975), 89 ff.

54. "Beschluß über die Grundsätze für die Gestaltung des Auftragleitersystems für wirtschaftlich entscheidende Aufgaben", GBl, T II, Nr. 27 (1970), 197-199; R. Gericke and W. Heerdegen, "Probleme der Anwendung des Auftragleitersystems in Forschung und Entwicklung", Die Technik, Nr. 10 (1970), 641; H. Wambutt, "Erfahrungen bei der sozialistischen Wissenschaftsorganisation in unserer chemischen Industrie", Einheit, Nr. 8 (1970), 1039.

55. Another factor obstructing the transfer of manpower was the housing shortage.

56. L.I. Breshnew, Rechenschaftsbericht des Zentralkomitees der KPdSU an den XXIV. Parteitag der KPdSU (Moskau: APN-Verlag/ Berlin: Dietz Verlag, 1971), 93.

57. "Höchstleistungen und sozialistische Großforschung", Die Wirtschaft, Nr. 8 (1969), 2.

58. K. Marx, Das Kapital. Dritter Band, in K. Marx and F. Engels, Werke, Band 25 (Berlin: Dietz Verlag, 1973), 89. See also 251, 276, 452, 688, 886, 892.

59. K. Marx, Das Kapital. Erster Band, in K. Marx and F. Engels, Werke, Band 23 (Berlin: Dietz Verlag, 1972), 654.

60. W.I. Lenin, "Was sind die Volksfreunde", in Werke, Band 1 (Berlin: Dietz Verlag, 1971), 169.

61. W.I. Lenin, "Entwurf eines Plans wissenschaftlich-technischer Arbeiten", in Werke, Band 27 (Berlin: Dietz Verlag, 1972), 312.

62. Ibid., "Ursprünglicher Entwurf des Artikels 'Die nächsten Aufgaben der Sowjetmacht'", 202.

63. C. Stern, "Abbruch eines Denkmals", Zeit Magazin, Nr. 27 (Juni 1973), 2.

64. D. Wahl, "Arbeit im Kollektiv - Quelle der Produktivität", Neues Deutschland (19. April 1969), 10.

65. G. Block, "Tendenzen in der Forschungsorganisation der Industrie", Die Technik, Nr. 5 (1969), 301.

66. Ulbricht, op. cit., note 5, 18.

67. Ibid.

68. L. Rouscik, "Die optimale Betriebsgröße", Die Wirtschaft, Beilage zur Ausgabe Nr. 5 (1968), 9.

69. Ibid., 7.

70. Ibid., 5.

71. Ibid., 8.

72. M. Kuczmera, "Optimale Betriebsgröße und Arbeitsproduktivität in den Industriezweigen der DDR", Die Wirtschaft, Beilage zur Ausgabe Nr. 5 (1968), 12.

73. Die Redaktion, Die Wirtschaft, Nr. 14 (1968), 16.

74. K. Hartmann, "Zur sozialistischen Konzentration und Kooperation", Die Wirtschaft, Nr. 14 (1968), 16.

75. Referat von Walter Ulbricht anläßlich der internationalen wissenschaftlichen Session zum 100. Jahrestag der Veröffentlichung des ersten Bandes des "Kapitals" von Karl Marx. Quoted by Hartmann, op. cit., note 74, 17.

76. G. Kuciak, "Konzentration und optimale Betriebsgröße", Die Wirtschaft, Nr. 16 (1968), 13.

77. See "Verordnung über die Bildung und Rechtsstellung von volkseigenen Kombinaten" and "Verordnung über das Verfahren der Gründung und Zusammenlegung von volkseigenen Betrieben", GBl, T II, Nr. 121 (1968), 963 and 966.

78. K.H. Behrendt, "Der ökonomische Nutzeffekt bei der Bildung volkseigener Kombinate", Sozialistische Finanzwirtschaft, Nr. 12 (1970), 11; and Nr. 13 (1970), 30.

79. Autorenkollektiv, Die Intensivierung der sozialistischen Industrieproduktion und die wachsende Rolle der Arbeiterklasse (Berlin: Dietz Verlag, 1975), 58-59.

80. Wikarski, op. cit., note 37, 385.

81. Ulbricht, op. cit., note 5, 19.

82. A. Lange, "Einige Probleme der sozialistischen Wissenschaftsorganisation", Wirtschaftswissenschaft, Nr. 9 (1969), 1282.

83. "Strategie der Großforschungszentren", Die Wirtschaft, Nr. 38 (1970), 6.

84. W. Ulbricht, "Für einen großen Aufschwung bei der Durchführung der Wissenschaftsorganisation in der chemischen Industrie", Die Wirtschaft, Beilage Nr. 11 zur Ausgabe Nr. 14 (1970), 13.

85. G. Prey, "Zur Entwicklung der Großforschung im allgemeinen und der Aufbau der Großforschung im besonderen", Technische Gemeinschaft, Nr. 7 (1970), 8.

86. See Chapter Five.

87. Lorenz and Haker, op. cit., note 29, 5.

88. H. Pöschel, "Ideologische Probleme der Leitung wissenschaftlich-technischer Arbeit", Einheit, Nr. 10 (1972), 1281.

89. "Die Durchführung des Volkswirtschaftsplanes im Jahre 1970", Neues Deutschland (11. Juni 1970), 6.

90. "Aus dem Bericht des Politbüros an die 14. Tagung des ZK der SED", Neues Deutschland (10. Dezember 1970), 4-8.

91. See for instance: M. Keren, "The New Economic System in the GDR: An Obituary", Soviet Studies, No. 4 (1972/1973), 554; G. Leptin and M. Melzer, Economic Reforms in East German Industry (London: Oxford University Press, 1978), 78; and Deutsches Institut für Wirtschaftsforschung Berlin, Handbuch

DDR-Wirtschaft (Reinbek bei Hamburg: Rowohl Taschenbuch Verlag, 1977), 61.

92. "Die historische Leistung W.I. Lenins auf dem Gebiet der Wirtschaftstheorie und Wirtschaftspolitik", Die Wirtschaft, Beilage zur Ausgabe Nr. 45 (1969), 5.

7
The Effort to Link Academic Science with Industry, 1945–1975

POLICY MEASURES

> The victorious working class of the German Democratic Republic, under the leadership of the Socialist Unity Party in alliance with the Soviet Union, enables the Academy of Sciences of the GDR to work creatively in the interests of the people for the first time in its history. It therefore fulfils the legacy of its founder Gottfried Wilhelm Leibniz.[1]

After the Second World War, the Academy of Sciences was converted from an élite learned society to a scientific research academy, and in higher education the former preserves of the "privileged bourgoisie" opened their doors to persons from "worker and peasant" backgrounds.[2] The party leadership requested that more academic work be conducted and exploited in areas of importance to the economy and recommended closer connections between academic institutions and industry, stating that besides increasing the stock of scientific knowledge more work should be done on improving the quality of industrial products and on raising labour productivity.[3]

The onset of the Cold War was reflected in efforts to orientate GDR science away from West Germany and towards the Soviet Union. Visits to Western conferences were discouraged and warnings were given about espionage and diversion centres" in West Berlin, whose aim amongst other things was to disrupt scientific work in the GDR. The party called for close contacts between academic institutions in the GDR and USSR, and invited Soviet academics to speak about the organisation of scientific work in the Soviet Union.

> During the course of our five-year plan, we have received substantial scientific assistance in the most varied of areas from Soviet scientists. Some of our scientists, engineers and technologists realised for the first time how far behind we are in many fields .... The genius of science, J. Stalin, spoke in 1938 as follows about the progressive nature of science ...[4]

There was a significant expansion of scientific-technical capacity in the academic sector. From 1946 the Academy acquired for the first time in its history research personnel and research establishments. Manpower was attracted from industry, and research establishments were either taken over from the Kaiser-Wilhelm-Gesellschaft and industry or were newly founded. Table A.12 in the Appendix shows that fewer than 1,000 persons were employed in the Academy in 1949, more than 8,000 in 1960, and more than 16,000 in 1975. It is known that this expansion was due primarily to the promotion of science and technology. The bias towards science and technology in the Academy was also reflected in the composition of its membership: between 1946 and 1960 the balance of members in the natural sciences on the one hand and humanities and social sciences on the other changed from 1 : 1 to 2.2 : 1.[5] Higher education was orientated more strongly towards the needs of the economy. On Soviet occupied territory in 1945 there were eight university-type institutions, two of which were technical; in the Western Zone there were eleven technical universities. By 1960 the GDR had 44 institutions having university status; many were specialised and ten were technical. Table A.13 shows that whereas the GDR had a slightly higher proportion of university students graduating in mathematics and natural sciences than the FRG (7.4 % : 5.5 % in 1967), its proportion in the technical sciences was considerably higher (20.4 % : 9.3 % in 1967). The GDR also had higher proportions than the FRG for medicine and agricultural science, but relatively low proportions for the humanities, social sciences and art.

Measures to improve contact between academic science and industry included the encouragement of academic consultation work; of staff exchanges and joint meetings between the two sides; and of students undertaking projects of industrial relevance.[6] In 1962/63, the Academy was given the task of planning basic research and coordinating it with industry.[7] With the introduction of the New Economic System, increased emphasis was also placed on academic science conducting contract research for industry.[8] This, according to official propaganda, was to assist science to become a "direct productive force".

Prior to the NES the amount of contract research in the Academy had apparently been very low. The ratio of contract finance to the Academy's total wage fund in 1960 was 3.2 %; in 1961, 2.1 %; and in 1962, about 1.8 %.[9] For the higher education sector, State Secretary Gießmann stated that about 40 % of the research funds in 1963 came from enterprise and VVB contracts. The figure for the technical sciences was about 60 %.[10] If these statistics can be trusted, then university research in the GDR was heavily dependent on industrial finance even at the inception of the NES, and the dependence appears to have been much higher than that in the OECD countries: in 1963/64 the OECD countries having the highest proportions of business enterprise financed R&D in the higher education sector were Spain and Ireland with 6.8 % and 5.1 % respectively.[11] The fragmentary data on contract finance do not reflect all the research conducted for industry in the academic sector. Both the Academy and the universities received funds from the state budget for topics of direct relevance to industry. Gießmann stated in 1963 that 30 % of the finance provided by the central agencies was intended for practical application.[12] Professor Beckert, rector of the Technical University "Otto von Guericke", Magdeburg, stated that since its founding in 1953 his university had maintained close contact with the heavy machine building industry in the area. Research tasks, diploma projects and dissertations had been completed on topics related to machine building, though most of the research had been financed from the state budget.[13]

After 1963, the level of contract finance in the Academy and universities increased, but detailed statistics are not available. One GDR academic, Pohlisch, reported that about 23 % of university research "capacity" in 1966 was bound by contracts. In 1967, the figure was almost 33 %.[14] He did not define what he meant by "capacity", but reference to another source suggests that it meant "full-time equivalent manpower".[15] His figures are therefore not comparable with those of Gießmann. The technical universities had the highest level of contract finance. The Technical University for Chemistry in Leuna-Merseburg, for example, devoted 60 % of its research "capacity" to contracts with the chemical industry in 1965.[16]

The party leadership, however, was not satisfied with the relationship between academic science and industry, and in 1967/68 ordered important reforms of the Academy and universities. The aim was to forge a more direct academic science-industry link, and the policy was part of the "offensive strategy for technological change" discussed in the previous chapter. According to Ulbricht it was a case "of linking the scientists of the Academy closely and effectively with socialist large-scale industry, mainly with the combines and large-scale research centres, so that together pioneer and supreme achievements can be made".[17] As in industry, there was much emphasis in the Academy and universities on forecasting work and the "concentration of research".

> The Academy reform was decided upon in the summer of 1968 to promote the development of the German Academy of Sciences into the research academy of the socialist society. This decision was based on forecasts about the most important areas of natural science and technology, and aimed at incorporating the research potential of the Academy complexly and with high effectiveness in the societal reproduction process of our country.[18]

The "concentration of research" was the remedy for what was considered to be too great an "institutional and thematic fragmentation" of academic research, and a consequent lack of efficiency. In the opinion of W. Hartke, president of the Academy from 1958 to 1968, socialist teamwork in research was essential: scientific knowledge and its technical application were growing exponentially, and it was imperative for a modern socialist industrial state to have a strong scientific potential at its disposal. He regretted the fact that up to the time of writing, 1968, socialist teamwork had not developed satisfactorily in the Academy.[19]

To stimulate the Academy and universities into performing research of interest to industry and relevant to "structure-determining areas", the global institutional finance of academic research from the state budget was abolished in January 1969 and replaced by contract, task-based financing. The Academy and universities now had to seek societal partners.[20] Pöschel commented on the measure as follows:

> Previous requests for an orientation of our scientific-technical work towards the needs of the national economy had the character of an appeal for the majority of our scientists, engineers and employees. The fulfilment or non-fulfilment of these requests as a rule had no directly felt consequences for our collectives. With the introduction of task-

> based financing, everyone will be forced each day to check anew whether the tasks and aims of scientific work coincide with societal interests.[21]

Ulbricht had stated in 1967 that if industry financed research projects in the academic sector, it would also make sure that the necessary provisions were made for transferring the results into production.[22]

Increased efforts were made in the higher education sector to align training with the needs of industry. Faculties and institutes were replaced by so-called "sections" to facilitate interdisciplinary work, curricula were revised, and the student was to be equipped with "knowledge of the latest discoveries, and methods of transferring results rapidly into production". According to one publication, the universities were to become "combines of science".[23] Ten new engineering universities were created in 1969 and aimed in general at being useful to industry and in particular at relieving the shortage of qualified "technologists" in design and production preparation. "Neither from their profile nor the available capacity can it be the task of the engineering universities to engage in purely theoretical research topics."[24] Between 1968 and 1970, the total number of university admittances in the technical sciences, mathematics, natural science and economic science doubled whereas the overall increase for admittances into all subjects was about 50 %.

Both during and after the "offensive strategy", propaganda was devoted to the "socialist teamwork" of scientists and engineers from an academic institute and an industrial enterprise.[25] Importance was also attached in the seventies to the creation of joint "Technika" and "Academy-Industry Complexes": the former were R&D facilities shared by the Academy, universities and industry; and the latter, joint facilities of the Academy and industry. No precise figures could be found in the literature, however, as to how many of these establishments were created. One source published in 1975 stated that "several" joint "Technika" would be in existence by 1980,[26] and another published in 1976 stated two "Academy-Industry Complexes" had already been founded.[27]

An important component of party policy to strengthen the link between academic science and industry during the period 1945-1975 involved the attempt to win over the academic community. Most of the scientists employed in the immediate postwar period were of "bourgeois" background, and adhered to the traditional German ideal of "freedom of teaching and research". An early example of outspoken academic criticism is that by Walter Friedrich, who was president of the Academy from 1951 to 1956:

> Above all, the Academy also feels that more scope should be granted to basic research and the personal initiative of the researcher. The planning of fundamental scientific work and the organisation of research should be freed of bureaucratic obstacles, such as are still often exercised by the Central Office for Research and Technology, and should be made more flexible. In spite of many conferences and all kinds of efforts, it must be emphasised that a qualitative change in this respect has not yet been achieved. In addition an often met underestimation of science and its importance should be replaced by a correct valuation. Similarly, non-competent agencies should stop patronising science. One should, on the contrary, make much better use of the scientists' rich experience and listen to their opinion.[28]

Academic criticism was less direct after 1961, as it was no longer possible to move across to the West, but passive resistance apparently continued. The planning of academic work, for instance, did not meet with much enthusiasm, and some scientists felt things would be better if the party kept to politics and left science to themselves. Even certain social scientists held such opinions.[29] Another issue was the so-called "world level" in science and technology: there were still researchers who continued to identify this concept with the "West level" and failed to recognise that the Soviet Union was becoming in ever-increasing measure the leading country of the world, not least with respect to science and technology.[30]

The party leadership attempted to deal with these problems in at least four ways. One was the granting of material privileges to the scientific-technical intelligentsia; the latter received higher incomes than most members of the population, longer holidays and preference in accommodation. A second way was by inserting trusted party members into key positions: in accordance with the principle of "democratic centralism", they had to try and ensure that the will of the leadership was done. A third was the "ideological help and re-education" of the academic community, particularly by the party organisation.[31] A fourth was the training of a new socialist intelligentsia. By the late sixties and early seventies, a large proportion of the academic community had been brought up and trained in the GDR.

The following sections look in more detail at aspects of the research link between academic science and industry. First, however, two points concerning definitions and theory should be made. The concept of "academic research" to a large degree overlaps but is not coincident with that of "basic research". Academic research is understood in this book to be research conducted by the Academy of Sciences and the establishments of higher education. Basic research may be defined as original investigation undertaken to gain new scientific knowledge, but not having any specific application. In the GDR basic research is conducted mainly, though not entirely, in the Academy and universities: some is performed in industry. Moreover the Academy and universities conduct some applied research and development. Whilst most people would probably subscribe to the above definition of basic research, they would differ when actually having to categorise a particular piece of research: the same research project might be designated "basic" in the Academy context, and "applied" in an industrial one. Western empirical investigations into the role of basic research in technological change have reached very different conclusions. The study "Hindsight" found that basic research played no noteworthy role, contributing only 0.3 % of all the R&D events identified in the innovation sample.[32] "Traces", by contrast, found that 70 % of key R&D events were non-mission oriented. Whereas "Hindsight" stated that only 9 % of all R&D events came from the universities, the corresponding figure in "Traces" was 60 %.[33] Other studies have emphasised the contribution of basic research to innovation as being "latent" rather than direct.[34] The conclusions of these investigations depend critically upon factors such as definitions; the nature of the innovation sample; the method in which innovations are dissected into R&D events; the length of the time period investigated; and possibly also the subjective desire to "prove" what is deemed desirable. In this book it is assumed that basic research is important for technological change both in a direct sense, i.e. the use of discoveries; and indirectly, through "problem solving", the training of manpower, and the assimilation of foreign technology.

There were at least two necessary conditions for an effective link between academic research and industry: the command of a reasonably strong academic research potential, and the existence and utilisation of suitable channels of communication between academic science and industry. These topics are explored in the next two sections. The last section deals with the far-reaching measures to couple academic research with industry in the context of the "offensive strategy".

ACADEMIC RESEARCH MANPOWER AND FACILITIES

Detailed data on the number of qualified scientists and engineers in the academic sector are not available but it is known that in 1970, more than 15,000 persons were engaged on R&D work in the Academy and universities. The research effort in the academic sector was roughly one-fifth the size of that in industry, and constituted about one-tenth of the GDR's total R&D effort.[35] Most of this academic research effort was located in the Academy of Sciences.[36]

Information for the Academy suggests that the expansion of scientific personnel was not well balanced with regard to the proportions of technical support staff and manpower based in prototype construction and testing. For effective connections between academic science and production, it was equally as important for the Academy to have some persons qualified and deployed in development work as it was for industry to have some staff conversant with the latest findings of basic research. Although the Academy certainly employed personnel qualified in development work, GDR sources suggest that most were not actually engaged in this kind of activity. Over the period 1965-1973, the Academy's scientific-technological manpower in the category "research and development" increased by 25 % compared with only a 5 % increase of manpower in the category "prototype construction, 'technika' and testing". The overall share of the latter category in total scientific-technological manpower in the Academy decreased from 5.1 % to 4.3 %.[37]

GDR sources complained that the ratio of technicians to scientists for the Academy was too low. The development of this ratio between 1950 and 1975 is given in Table 7.1 but a rough comparison with the FRG calls these complaints into question: in 1969 the ratio QSE : Technician for the "private non-profit" sector of the FRG was 1 : 1.3.[38] Nevertheless there is also evidence to indicate that, unlike the situation for the FRG, not all GDR workers in the statistical category "Technician" were qualified technicians.[39]

The Minister of Higher Education, H.-J. Böhme, cited the figures given in Table 7.2 for the number of university teachers and "collaborators" in 1971 and 1973. The scientific-technological research manpower in the universities was, of course, substantially lower than the number of personnel listed in the table, as the figures contain all subjects and not just natural science and technology, and they also include the time spent on teaching and administration.

Table 7.1
Ratio Scientist : Technician in the Academy

1950:	1 : 3.8
1960:	1 : 3.1
1971:	1 : 1.7
1973:	1 : 1.5
1975 (Plan):	1 : 1.3

Source: Calculated from data given in Table 3 of *Das Forschungspotential im Sozialismus* (Berlin: Akademie Verlag, 1977), 112.

Note:

This source differentiates between the following occupation categories for the Academy:
- Wissenschaftler (einschließlich wissenschaftsorganisatorisches Personal in der Forschung);
- Wissenschaftlich-technisches Personal;
- Verwaltungspersonal;
- Betriebs- und Betreuungspersonal.

The first category is represented in Table 7.1 by the term "scientist", and the second, by "technician".

Table 7.2
Teachers and Collaborators in GDR Universities 1971 and 1973

	University Teachers	Academic Collaborators*
1971:	4,962	17,523
1973:	5,083	18,211

Source: H.-J. Böhme, "Zum Stand der Erfüllung der Beschlüsse des VIII. Parteitages an den Universitäten und Hochschulen", *Das Hochschulwesen*, Nr. 9 (1973), 263.

*Wissenschaftliche Mitarbeiter

There were complaints that the proportion of time spent on research work in the universities was low.[40] Hager, who had held various high-level positions in science, education and propaganda, called in 1969 for a "continual struggle" against the often unproductive use of university scientists' "labour time fund" (i.e. working time).[41] Bode, of the section "socialist industrial economics" at the Technical University of Dresden, stated in 1974 that whereas research should take up about 30 % of the working time, investigations showed that the average proportion at his university was lower. The figure for professors was 23 % and for lecturers, 25 %, and only 20 % of the university lecturers could spend more than 30 % of their total working time on research.[42] Estimates indicate that universities in the FRG devoted a significantly greater proportion of total working time to R&D: for 1973 this proportion was put at 55 % in natural science and 50 % in engineering and technology.[43] The GDR's 1976 plan for the universities stipulated that one-third of the projected increase in research capacity was to be achieved through "opening up inner-reserves". The time actually spent on research was to be raised by cutting down administrative work and by using students' abilities more effectively.[44] A source published towards the end of 1976 claimed that the situation had improved: in most scientific and technical disciplines, research amounted to 30 % of the total working time.[45]

Concerning R&D facilities, the Academy and the universities had problems similar to industrial establishments. There were difficulties in obtaining supplies on time; modern scientific equipment was not easy to acquire; the technological level of research apparatus was criticised; and facilities for testing and preparing scientific results for production were considered inadequate. Probably the situation in the Academy was better than in the universities. Linnemann and Wykowski stated in 1976, for instance, that it was no longer possible for universities to have the most modern apparatus and equipment.[46] Yet in spite of the greater prestige of the Academy as a research institution, research provisions there were viewed as unsatisfactory. One survey conducted in the Academy asked researchers to classify the technological level of major pieces of apparatus in relation to their research needs. There were three possible classifications: "good", "satisfactory", and "inadequate". About 20 % of all the apparatus was judged to be "inadequate".[47] Computer facilities in the Academy were also criticised: whereas the Academy's electronic data processing capacity expanded rapidly in the seventies, and in 1977 its Centre for Computing had the largest computer facility in the GDR, capacity per R&D worker was reported to be considerably below that of other scientific institutions both at home and abroad.[48]

Measures to improve academic R&D facilities were adopted only belatedly: from the late sixties and particularly from 1975. There was an expansion of technological capacity in the form of "Technika", which permitted the construction of prototypes, small series, and the testing of technologies.[49] Hofmann in 1975 stated that roughly one-third of the Academy's scientific institutes had "Technika" and a further one-third was planned to have them between 1980 and 1985.[50] As already mentioned, joint "Technika" were created by academic institutions and industry. Scientific instrument building was strengthened not only in industry but in the academic sector itself. A "Centre for Scientific Instrument Building" was created in the Academy and had charge of a number of small instrument building enterprises or part of their production volume.[51] The production capacity of the Academy doubled between 1970 and 1975.[52] In 1975 the Academy covered 15 % of its needs with respect to scientific instruments, and by 1980 it was to cover 25 %.[53]

More intensive use of existing equipment was also called for. The Academy was set the task of increasing its use of expensive research apparatus from 48 hours per week in 1976 to 60 hours per week in 1980.[54] The universities had to improve on their "unsatisfactory" average "utilisation level" of scientific apparatus of 50 %.[55] It should be noted, however, that the index "utilisation level" has serious limitations for it depends upon factors such as the type of research, the degree to which the apparatus can be shared by other researchers, and the technological level and state of repair of the apparatus. An investigation at the Martin-Luther-University of Halle-Wittenberg revealed a number of reasons for 83 fixed assets not being used: some of the assets were out of date, some needed spare or foreign parts, some were not worth repairing, and others had been delivered incomplete or faulty.[56] To promote joint use of apparatus and facilities, new organisations such as "apparatus commissions", "methodical-diagnostic centres" and "central loan and maintenance stations" were set up,[57] and "cooperation councils" attempted to coordinate the facilities of research institutes and establishments in the same locality.[58] Nevertheless, the "egoism" of certain research collectives was criticised for obstructing such efforts.[59]

COMMUNICATION CHANNELS BETWEEN ACADEMIC SCIENCE AND INDUSTRY

Addressing research directors of the Academy in 1962, Hermann Klare alleged that whereas the Academy institutes had close connections with the universities, particularly through joint posts, relations with institutes in industry were often distant and sometimes non-existent.[60] In 1969 Klare, now president of the Academy, complained that partners in the academic and industrial sectors failed to decide in good time when and how transfer of scientific results into production should take place: arrangements which existed purely on paper were not sufficient.[61] Attempts to guarantee rapid transfer of results from the academic sphere to industry by coordinating plans of the enterprises with plans for science and technology of the Academy and the universities were not a conspicuous success.[62] Many Western scholars regard personal contacts and "person-embodied" knowledge to be the decisive mode of information transfer between science and technology,[63] and in this section five channels of "person-embodied" knowledge flow in the GDR will be considered: university graduates in industry; joint representation on committees; contact through "socialist teamwork" and joint facilities; temporary staff exchanges between academic science and industry; and job mobility between the two sectors.

Although new university graduates could bring up-to-date knowledge with them into industry and a capacity to follow new developments, management was not enthusiastic about employing them.[64] By the early sixties the party leadership was expressing serious concern about the shortage of graduates in industry, and in the succeeding period the situation improved: the number of graduates employed per 1,000 employees in 1961 was 6.0; in 1971, 17.0; and in 1975, 27.9.[65] Nevertheless in the early seventies there was apparently still a shortage of graduates in the enterprises. An investigation conducted in 1969-1970 into the poor performance of the "Furniture Combine Berlin", for example, found amongst other things that the management was not very well qualified and there was only a small number of graduates on the staff. Except for the production director, none of the departmental directors in the whole of the production division had an engineer's qualification. The R&D section did not have any graduates.[66] It was reported in 1972 that new graduates going into industry often had unrealistic ideas about the kind of work they were going to do. This is hardly surprising in view of the demands of the party for "pioneer achievements" and "structure-determining products".

> The graduates of our universities and technical colleges are often inclined to believe that when they are engaged in scientific work it must involve something new of principle. This attitude or tendency continues of course to a certain extent during their work in industry. In fact it is often the case that the economic gain is much greater for the further development of a known process ...[67]

The representation of industry on academic committees and vice versa constituted, in theory at least, an excellent means of exercising influence and making opportunities and needs known. In practice, up to 1968 the work of industry in the committees of higher education did not match the reciprocal contribution of the universities and colleges in the various industrial councils. Many enterprise representatives worked only "sporadically" in higher education committees, a notable contrast to the situation in the USA where the majority of university trustees were businessmen.[68]

When joint work in the form of transfer teams and "Technika" took place, it appears to have been successful. The fact that scientists and engineers from the academic sector and industry were working side by side meant that conditions for effective communication were far higher than via reports and plans, and discontinuities in projects could be more easily avoided. An important advantage which the GDR had over the USSR in this kind of work was the geographical proximity of academic science and industry, particularly through the greater specialisation of academic research establishments in subjects relevant to local industry.[69]

Geographical proximity was also an advantage in the promotion of temporary staff exchanges between the two sectors. Ideally, such exchanges served to deepen the awareness of each side about the abilities and needs of the other, and to set up a useful system of personal contacts. Yet proximity alone cannot guarantee exchanges and in practice, the academic sector showed greater interest in such exchanges than industry. At the Technical University for Chemistry at Leuna-Merseburg, 15.2 % of all university teachers of economic subjects had a temporary job in industry in 1970. So too had 18.8 % of all tenured "academic collaborators" in the sections "socialist enterprise management" and "cybernetics, mathematics and data processing".[70] On the other hand, the Academy received only a "weak response" from the industrial ministries to its offer of temporary places for enterprise employees. Klare reported in 1967 that a total of 245 places had been offered, though only one contract had been made between a VVB and an Academy institute.[71] No doubt the lack of graduates in industry was one reason for this. Krieger noted in 1970 that the three combines working a temporary exchange scheme with the Technical University for Chemistry in Leuna-Merseburg tended to place university visitors into management positions in priority areas, particularly in research and rationalisation. "In these areas there is a distinct shortage of specialists." As a rule staff were selected from other parts of the combine to spend some time at the university.[72]

Job mobility between academic science and industry is an important means of transferring results, and in the West it appeared to be highest in the United States and lowest in Japan; that in the FRG, the UK and France was somewhere between.[73] In the GDR job mobility did take place but it was mainly in one direction, from industry to the academic sector, and little seems to have been done to attract experienced qualified manpower from the academic sector to industry. Even when wages in industry were higher than in the academic sector, other factors such as pensions, sickness-benefits and prestige were less favourable.[74]

A fundamental difficulty with all the measures adopted to promote communication and contact between the academic sector and industry was that they could only be effective if both sides had sufficient interest in research collaboration and technological change. Over the period 1945-1975, with the possible exception of the years 1968-1970, the root problem was that the Academy and the universities were primarily interested in longer-term basic research tasks, whereas industry was more interested in output plan fulfilment and current difficulties of production. Representatives of industry often expressed the opinion that the content and direction of academic research was insufficiently concrete and flexible to be of use. Furthermore, cooperation partners in the academic sector could not always be trusted to guarantee the continuity of a project; sometimes research capabilities were transferred to other tasks, and the partner in industry left in the lurch.[75] Representatives of academic science said that most initiative for collaboration came from them rather than industry.[76] In 1966 Hartke stated in a presidential address to the Academy that enterprises were frequently only interested in cooperating with the Academy if it could offer developments ready for production.

> On the part of industry there is often a reluctance to accept the risk of a development together with the Academy. The responsibility for taking up new results of basic research in good time and further developing available research results to the production stage is quite regularly repudiated.[77]

THE OFFENSIVE STRATEGY

Implicit in this strategy was perhaps a somewhat oversimplified view of the relationship between academic science and industrial technology. A decision of the State Council on the Academy reform stated that:

> ... the research of the German Academy of Sciences in decisive areas is, beginning with the forecast and the conceptualisation of objectives, to orientate unremittingly towards priority development of the bases of completely new technologies and labour principles ... according to the principle of "overtake without catching up".[78]

Those who expressed doubts about radical technological change functioning in this manner, possibly envisaging a more complex relationship between academic science and technology, were "enlightened" by the party organisation. Ebel, the secretary of the party organisation at the Academy Research Centre in Leipzig, disputed in 1968 the view that "supreme achievements" were only to be expected every ten to fifteen years or so. Such a viewpoint hardly stimulated the "struggle for scientific top achievements"; every research collective had to appreciate the importance of this struggle, and a prerequisite for achieving such results was the use of prognostic thinking. As a rule, he thought, research collectives would be able to produce their "supreme achievements" before the end of the perspective plan period in 1975.[79]

Preoccupation with concentrating academic research was by no means a phenomenon specific to the GDR. In West Germany efforts were made from 1968 to achieve closer cooperation between the universities and independent institutes and to overcome institutional frontiers through the initiation of so-called "special research areas" (Sonderforschungsbereiche). Yet it is difficult to avoid the view that the concentration of research in the GDR was overdone: little regard seems to have been paid to the nature of the research involved and whether it actually called for large facilities.

Between 1968 and 1970 the number of institutional units in the Academy was reportedly reduced by about two-thirds. The development of "larger and more efficient central institutes" employing between 200 and 800 persons was viewed as being a major precondition for the attainment of "pioneer achievements". Large, interdisciplinary, problem-oriented research teams were created comprising between 50 and 80 scientists, and sometimes more than 100.[80] Similar developments took place in the universities. Both Academy and university institutes were expected to engage in "large-scale research" cooperation with R&D establishments and production plants of industry.[81] Three-quarters of the research potential at Jena University was oriented towards scientific instrument building in connection with the firm "Carl Zeiß Jena". From 1968 the Technical University "Otto von Guericke" at Magdeburg specialised in training and research for heavy machine and plant building.

GDR publications suggest opposition to the policy of concentrating academic research. Large-scale teamwork and cooperation were hindered by "institute egoism" and the reluctance of scientists to change their research area, join a larger collective, and possibly to move to a different institute.[82] Hermann Klare stated:

> ... an extensive amount of persuasion work had to be done in order to stop research on tasks which individual scientists considered important from their personal and sometimes even objectively substantiated point of view, in favour of other tasks more important for society .... Discussions concerning these ideas and tendencies were conducted with the requisite understanding, though with the necessary determination.[83]

In all likelihood many academic scientists managed to continue working as before simply by paying increased lip-service to the utility of research, the imperative of large-scale operations, and the necessity of "pioneer achievements". Ebel noted, for instance, that some research managers tried to avoid "concentrating" research simply by relabelling old research topics with a new "pioneer achievement" name.[84] And Professor Liebscher of the Technical University of Dresden pointed out that the newly formed university "sections" could in fact be subdivided into "areas, work groups etc.".[85]

It would be beyond the scope of this chapter to undertake a comprehensive analysis of the system of contract, task-based finance operating in the academic sector from 1969. Typically, the Academy complained that industry proposed contracts which were poorly formulated and not properly thought out.[86] Sometimes contracts for the current plan-year were only finalised in the third quarter.[87] And there was the perennial difficulty of how to calculate the economic use of the results provided. Perhaps a more fundamental question is whether the party leadership's aim of establishing such intimate links between academic science and industry through contracts was appropriate. How justified was the concern voiced from sections of the academic community that extensive contract financing involved a subordination of academic science to industry and a neglect of basic research?

This concern was not new. Early publications indicate that the party leadership often took some trouble to reassure the academic scientists that basic research would not be sacrificed for short-term gains.[88] In 1968/69, however, the situation was more acute. Ruhle warned that the position of the GDR as a developed industrialised country would be endangered if no energetic efforts were made in basic research. It was possible, he said, in the short term to orientate mainly towards applied research and development, but in the long term a neglect of basic research would lead to a deceleration in the rate of development of indigenous science and technology. The fact that most findings of basic research were published did not mean that one could manage without a domestic effort in basic research. A domestic capability created both a reserve of qualified personnel for applied R&D and built the basis for effective usage of international findings.[89] Gripinski warned that universities ought not to become specific "industrial ministry universities": they had responsibilities for the whole of the Republic and for Comecon.[90] In 1970, Lauter, the general secretary of the Academy, complained that the Academy was doing "direct industrial research".[91] Similarly, Linnemann and Wykowski for the universities stated at a later date that enterprises had often approached the universities with work which went far beyond topics in basic or applied research: it was almost pure production work.[92]

To put such warnings and criticisms into some kind of perspective, it is useful to consider firstly the extent to which academic science was financially dependent on industry, and secondly the level of basic research. The Academy of Sciences was most dependent on industry for finance in 1971: in that year industry provided 56 % of the Academy's financial resources for R&D.[93] This appears to have been a much higher proportion of industrially financed academic research than that evident in the OECD countries. In 1971 the OECD's "private non-profit sector" had the highest level of business enterprise finance of research, and was followed by the "government sector". Within the "PNP sector" only Japan approached the level of the GDR with 54 % of industrial finance; then came the UK with 37 %, and Sweden with 21 %. In the FRG such finance in the "PNP", "government", and "higher education" sectors was very low, accounting for 1.8 % of total research finance in the first sector, 0.2 % in the second, and 3.7 % in the third.[94] It must be remembered, however, that the statistical conventions used by the GDR and OECD are not identical and this probably acts to exaggerate the relative level of dependence of academic research upon industry for the GDR: the GDR would consider finance from industrial ministries as coming from "industry", whereas the OECD would regard this finance as stemming from "government". There was a similarity between the two Germanies regarding the areas of academic research financed by industry. Table A.14 contains an experimental comparison of industrial contracts to the Academy in the GDR, and to universities and research institutes (of Bund and Länder) in the FRG. It shows that in both countries, the three main sectors were chemicals, electrotechnology, and machine building.[95]

As noted in the first part of this chapter, the higher education sector appears to have had a high proportion of research financed by industry even before the introduction of blanket contract financing in January 1969. Scherzinger writes that the "contract research quota" rose from about one-third to 80 % between 1967 and 1971, and assumes this quota is the proportion of total research finance received through contracts with industry.[96] In fact the figure for 1967 relates to research "capacity", which is probably full-time equivalent manpower, whereas the figure for 1971 relates to research "tasks".[97] Moreover, industry was not the only contract-awarding body in the national economy. What can be said is that the literature gives an impression of a high proportion of university research being financed by industry, and the level was probably very much higher than in the West, even allowing for differences of statistical definition.[98]

Thus whilst a convincing case can be made for industrial influence in the academic sphere, it is doubtful whether the policy of heavy financial dependence of the Academy and universities on industry was a good thing. The complaints about a subordination of academic science to industry had foundation, as there was definitely a danger of the Academy and universities concentrating too much on short-term applied projects for industry at the expense of longer-term fundamental work.

To what extent was basic research being "neglected"? This question is easier to pose than to answer, but one might begin by differentiating between the basic research conducted in the academic sector and that conducted within industry. In the Western countries, the contribution of industry to basic research has been quite substantial. Considering basic research in the business enterprise sector as a percentage of all basic research, figures for 1971 were: West Germany 18.9 %; France 10.4 % and the USA 15.6 %.[99] In West Germany, basic research accounted for 7.3 % of all business enterprise R&D in the manufacturing sector in 1971; in Japan the figure was 9.4 %; in the UK (1969), 4.0 %; and in France and the USA, 3.4 %.[100] No statistics are available on the level of basic research in GDR industry, though sources suggest that it was fairly low. Ruhle complained in 1968 about the "insufficient" level of basic research in industry, and it was reported in the early seventies that this level had "dropped".[101] It is most unlikely that the level of basic research in GDR industry exceeded that in the industries of the OECD countries.

Concerning the academic sector, not all contract work was necessarily applied research. Table 7.3 shows that a total of 44 % of the Academy's income for R&D came from sources other than industry, and that 27 % came from the Ministry for Science and Technology. These sources may have awarded contracts for basic research. Moreover, even with industrial contracts there may have been scope for financing

Table 7.3
Sources of Finance for R&D in the Academy in 1971

Ministry for Science and Technology	27.2	
Users:		
Industry	56.4	
Ministry of Construction	1.5	
Ministry of Agriculture/Forestry/Food	4.3	
Other Areas	10.6	
of which Ministry of Health		5.5
	100	

Source: Autorenkollektiv, Das Forschungspotential im Sozialismus (Berlin: Akademie Verlag, 1977), 156.

basic research. Kramer and Linden emphasised in 1969 that cooperation between the Martin-Luther-University and the "VVB Agricultural Chemistry" should not be limited to applied research; basic and exploratory research should be included, and the university ought to remain a cultural centre.[102] Such scope for negotiation presumably depended on how prestigious the institute was and whether it was in a position to discriminate between potential contract donors. That some institutes were able to pick and choose is indicated by industrial enterprises not being able to contract out important work because their potential university partners were fully booked up.[103] A further indication is that certain institutes were able to abuse the bonus system for contract work by demanding the highest possible "performance-related surcharge" from prospective donors, or refusing to cooperate.[104] The fact that room for manoeuvre existed, however, does not necessarily mean that it was always used for the benefit of basic research. Grote in 1971 criticised the way that a number of directors and scientists of the Academy had accepted contracts from industry without trying to relate them to the Academy's longer-term tasks.[105]

To assert a "neglect of basic research" is to imply that there exist some optimal proportions for expenditure on basic research, applied research and development. But this is not the case: the best that can be done is tentatively to compare figures for the GDR with those for other countries, bearing in mind the problems of classification. Table 7.4 collects together some data on the distribution of R&D expenditure in the GDR between 1967 and 1972. If the data are to be believed, there was a considerable decrease in the proportion of expenditure devoted to basic research between 1967 and 1972. The figure for 1972 was certainly rather low compared with other countries, as may be seen from Table 3.4 in Chapter Three. Only Ireland, the UK and USSR had lower proportions whereas the FRG had twice the GDR's figure. Had GDR statistics been available in 1971, its comparative level of basic research might have looked even less favourable, for in that year the level of industrial financing was at its height. Furthermore, much of the GDR's "basic research" was probably "oriented" basic research, and this kind of work may have been classified by other countries under the rubric "applied research".

It would seem then that the party leadership accorded low priority to basic research and ironically, this contradicted the aim of achieving new "pioneer" technological changes, for it is the case that such

Table 7.4
Allocation of R&D Expenditure in GDR between 1967 and 1972 (%)

	Basic Research	Applied Research	Development
1967[a]	About 20		
1969[b]	19.9	27.8	55.3
1971[b]			65.8
1972[c]	13.5	30.2	52.0

Sources:

a) Interview with G. Prey, "Prognostik, Planung und Leitung auf dem Gebiet von Wissenschaft und Technik", Einheit, Nr. 10-11 (1967), 1361.
b) A. Lange, "Den Zyklus 'Wissenschaft-Technik-Produktion' beherrschen (Teil I)", Die Technik, Nr. 1 (1973), 8.
c) Autorenkollektiv, Das Forschungspotential im Sozialismus (Berlin: Akademie Verlag, 1977), 88.

advances often depend more on basic research than less radical ones. In spite of criticisms and cautious suggestions for a return to the central financing of basic research,[106] contracts with industry and other users provided most of the funding for academic research between 1969 and 1972. In 1972 a law was issued which in effect strengthened the position of basic research and weakened the dependence of the Academy and universities on industry.[107] Academic scientists were no longer totally dependent upon finance through contracts with users or the Ministry for Science and Technology. Planning competence for basic research was transferred from this ministry to the Academy and to the Ministry of Higher Education. Funding for basic research ran from the state budget via the Academy President and the Minister of Higher Education who acted as contract donors. Table 7.5 reflects the decreasing dependence of the Academy on user contracts.

This change of policy may not solely be attributed to the views and pressures from sections of the academic community: an important factor was probably the party leadership's loss of confidence in their own "offensive strategy". Publications now emphasised the importance of basic research. Hager, an important ideologist of the party, stated in 1972 that "a science which is only oriented to the needs of the next

Table 7.5
User Contracts to the Academy 1970-1975 as Percentage of R&D Finance

	1970	1971	1972	1973	1974	1975
User Contracts	71.0	72.8	62.7	50.8	56.6	48.2
of which Industry	51.2	56.4	46.9	34.9	29.2	31.3

Source: Autorenkollektiv, Das Forschungspotential im Sozialismus (Berlin: Akademie Verlag, 1977), 156.

few years would be ... just as inappropriate for socialism as a science which is exclusively oriented towards the future".[108] In 1974 he said that the promotion of basic research was an especial concern of the party's science policy. The party had repeatedly warned against any neglect of basic research and had stressed the responsibility of the Academy and university for this kind of research.[109] Hermann Klare characterised basic research as "the mother of tomorrow's production"[110] and Weiz, of the Ministry for Science and Technology, called upon industry to conduct more basic research.[111] Prey of the same ministry made the following comments to the Chemical Society in 1972:

> The use of silicates is at present to be found in the manufacture of glass, ceramics, building material, washing powder and other products. Our research, however, is currently almost totally oriented to products, to their manufacture and their behaviour in further use. Insufficient knowledge is available on the other hand about the behaviour of silicates under chemical reaction, about their properties. Some years ago we deployed a certain research potential, which then focused almost exclusively on the utilisation of the results in praxis. That was right. Yet in order to assure our forerunning position, we must now devote greater attention to basic research. In this way we shall create a basis for the later application of the results of silicate chemistry.[112]

One unrealistic demand was that academic scientists both point out the potential applications of their findings and personally fight for their realisation.

"Proceeding from the fundamental results of basic research, they must indicate in selected examples the use of the results in the national economy, and struggle for their transfer into the science-production chain as a new technology or new product."[113]

Publications in the seventies also indicated lower interest in the "concentration of research". This term appeared less frequently and when it did appear, the accent was different from that of the period 1968-1971. In an article published in 1972, for example, Kannengießer and Meske of the Institute for the Theory and Organisation of Science at the Academy stated that the concentration of research was necessary under socialist conditions, but could not and would not take place through the "administrative reprofiling of establishments, research groups and personnel", or even the release of personnel from research activity. Such measures were typical of conditions under capitalism.[114] No longer did GDR authors place especial emphasis upon the necessity of forming huge research teams and institutes. Instead they underlined the importance of project selection satisfying scientific and national-economic requirements; of efficient utilisation of "material and personal capacities"; and of research cooperation in the GDR and within Comecon.[115] Research cooperation had the advantage of flexibility: researchers could be grouped together fairly quickly to tackle a particular problem and once the task was completed they could disband and possibly regroup in a different combination.

Despite the various efforts to link academic science with industry from 1945 up to the abandonment of the offensive strategy, there were still familiar complaints in the seventies about the difficulties of research collaboration. J. Müller et al. stated in 1974 that the industrial enterprise often did not know which university department it should contact for prospective work: universities ought to publicise their interests and facilities.[116] Wykowski in 1976 pointed out that a decisive reason for "loss of effectiveness in transfer" was the insufficient attention paid by academic research collectives to the actual needs of their contract donor.[117] According to K. Müller, corresponding member of the Academy and director of the research centre at "Carl Zeiß Jena":

> Problems do not arise primarily from the fact that the natural scientist belongs to the institute, the laboratory scientist to the industrial

> R&D establishment, and the technologist directly to the production department. Rather, they result mainly because proper notice is not taken of the right and left hand neighbours in this chain. Seen from the beginning of the chain, the end disappears into the apparent chaos of operative enterprise activity. Conversely, seen from the end of the chain, the beginning dissolves into the mist of academic erudition. Adding the ideological problem that the work of this or that phase is looked down upon ... or one's own job considered more worthy than others, then one has the basis for manifold difficulties.[118]

Governmental encouragement of channels of communication involving person-to-person contacts were necessary and useful but could not replace the initiative of the two sides concerned. The 1972 law on funding academic research strengthened and shielded the position of basic research whilst maintaining the incentive for academic science to collaborate with industry: basic research was now funded "in principle" from the state budget but applied and possibly oriented basic research continued to be funded from "societal contract donors". The greatest problem concerned the low level of industrial interest in collaborating with academic science: what government policy lacked most of all was the provision of a real incentive for industry to pursue technological change vigorously.

NOTES

1. H. Klare and W. Hartkopf, _Die Akademie der Wissenschaften der DDR. Zum 275. Jahrestag der Gründung der Akademie_ (Berlin, 1975), 26. See also: W. Hartkopf and G. Dunken, _Von der Brandenburgischen Sozietät der Wissenschaften zur Deutschen Akademie der Wissenschaften zu Berlin_ (Berlin: Akademie Verlag, 1967).

2. E. Schwertner and A. Kempke, _Zur Wissenschafts- und Hochschulpolitik der SED (1945/46-1966)_ (Berlin: Dietz Verlag, 1967).

3. See e.g. "Weitere Entwicklung der Naturwissenschaften", _Nationalzeitung_ (13. März 1955), 2; and "Empfehlung zur weiteren Entwicklung und Verbesserung der Arbeit der Deutschen Akademie der Wissenschaften zu Berlin", _Mitteilungsblatt der Deutschen Akademie der Wissenschaften zu Berlin_, Nr. 4-5 (1955), 1-27.

4. W. Ulbricht, "Aufgaben der deutschen Wissenschaft bei der Schaffung der Grundlagen des Sozialismus", _Tägliche Rundschau_, Nr. 21 (25. Januar 1953), 5.

5. A.M. Hanhardt, Jr., "Die Ordentlichen Mitglieder der Deutschen Akademie der Wissenschaften zur Berlin 1945-1961", in P.C. Ludz (ed.), Studien und Materialien zur Soziologie der DDR (Köln und Opladen: Westdeutscher Verlag, 1964), 241.

6. See e.g. R. Müller and Bulla, "Über die Zusammenarbeit des Institutes für Elasto- und Plastomechanik der TU Dresden mit der Industrie", Die Technik, Nr. 2 (1963), 69-72.

7. "Beschluß des Ministerrates über Rolle, Aufgaben und die weitere Entwicklung der Deutschen Akademie der Wissenschaften zu Berlin vom 27. Juni 1963", Spektrum, Nr. 11/12 (1963), 374-375.

8. See in this respect also: "Rahmenverträge - neue Form effektiver Zusammenarbeit von VVB und Hochschule", Das Hochschulwesen, Nr. 10 (1965), 647-655.

9. H. Klare, "Probleme und Maßnahmen für die weitere Entwicklung der Institute der Forschungsgemeinschaft", Spektrum, Nr. 4 (1962), 169. In 1962, 25 of the Academy's institutes in the natural, technical and medical sciences had 107 contracts with 72 enterprises; the following year 32 institutes had 169 contracts with 107 enterprises. H. Wittbrodt and B. Gieltowsky, "Zur Entwicklung der Zusammenarbeit zwischen der Forschungsgemeinschaft und der Industrie im Jahre 1963", Spektrum, Nr. 5 (1964), 190.

10. E.-J. Gießmann, "Bildungswesen bestimmt maßgeblich Entwicklungstempo", in W. Ulbricht, Das neue ökonomische System der Planung und Leitung der Volkswirtschaft in der Praxis (Berlin: Dietz Verlag, 1963), 242.

11. A Study of Resources Devoted to R&D in OECD Member Countries in 1963/64. Statistical Tables and Notes, Vol. 2 (Paris: OECD, 1968), 362-363.

12. Gießmann, op. cit., note 10.

13. W. Sydow (ed.), Ideen - Projekte - Produktion. Aktuelle Fragen in Forschung und Entwicklung (Berlin: Verlag Die Wirtschaft, 1969), 232-233.

14. F. Pohlisch, "Bilanz eines Jahres", Das Hochschulwesen, Nr. 2 (1968), 99.

15. "Anordnung über die Planung, Leitung und Finanzierung von wissenschaftlich-technischen Aufgaben der Universitäten, Hoch- und Fachschulen vom 24. Januar 1969", GBl, T II, Nr. 16 (1969), 118. This source states that "research capacity" was the "labour time fund" (i.e. manpower) reckoned in terms of full-time equivalent units.

16. "Forschungskapazität vertraglich gebunden", Die Wirtschaft, Nr. 16 (1965), 4.

17. Die Deutsche Akademie der Wissenschaften auf dem Weg zur Forschungsakademie der sozialistischen Gesellschaft, Schriftenreihe des Staatsrates, Heft 12 (Berlin: Staatsverlag der Deutschen Demokratischen Republik, 1970), 104 and 119. For the universities see e.g. H. Wöltge, "Für eine dauerhafte Verbindung zwischen Hochschule und sozialistischer Praxis", Die Wirtschaft, Nr. 16 (1968), 5.

18. W. Ulbricht, "Akademie und Großindustrie im Kampf um Pionierleistungen", Neues Deutschland (14. März 1970), 3.

Also: W. Ulbricht, "Spitzenleistungen demonstrieren unsere ökonomische und politische Kraft", Die Wirtschaft, Beilage Nr. 9 zur Ausgabe Nr. 13 (26. März 1970), 14-16.

19. W. Hartke, "Hauptversammlung der Akademie", in Jahrbuch der Deutschen Akademie der Wissenschaften zu Berlin (1967), 125 ff.

20. GBl, T II, Nr. 16 (1969), 117 ff. See also: "Anordnung über die auftragsgebundene Finanzierung wissenschaftlich-technischer Aufgaben und die Bildung und Verwendung des Fonds Wissenschaft und Technik vom 30. September 1968", GBl, T II, Nr. 110 (1968), 859.

21. H. Pöschel, "Ökonomie und Ideologie in Forschung und Technik", Einheit, Nr. 8 (1969), 931.

22. W. Ulbricht, Zum ökonomischen System des Sozialismus in der DDR (Berlin, 1968), 552-553.

23. K. Hager, "Hochschulen werden zu Kombinaten der Wissenschaft", Die Wirtschaft, Nr. 15 (1969), 4.

24. E. Buzmann, "Forschungsarbeit und Praxisverbindungen aus der Sicht der Ingenieurhochschule", Das Hochschulwesen, Nr. 8 (1974), 250.

25. See e.g. "Die Deutsche Akademie der Wissenschaften auf dem Weg...", op. cit., note 17, 117; H. Groschupf, "Wissenschaft und Hochschulbildung im Dienste des gesellschaftlichen Fortschritts", Das Hochschulwesen, Nr. 10 (1974), 294.

26. "Akademie, Hochschulen und Industrie gründen Technika", Neues Deutschland (17. Dezember 1975), 4.

27. "Akademie-Industrie-Komplexe gegründet", Spektrum, Nr. 7 (1976), 33.

28. W. Friedrich, "Interview der 'Wirtschaft' mit dem Ersten Vertreter des Präsidenten der Deutschen Akademie der Wissenschaften zu Berlin", Mitteilungsblatt für die Mitarbeiter der Deutschen Akademie der Wissenschaften zu Berlin, Nr. 1-4 (1956), 32.

29. G. Heyden, A. Kosing, O. Reinhold and H. Ullrich (Redaktion), Sozialismus-Wissenschaft-Produktion. Über die Rolle der Wissenschaft beim umfassenden Aufbau des Sozialismus in der DDR (Berlin: Dietz Verlag, 1963), 32.

30. H. Ebel, "Die Akademiereform verlangt politisch-ideologische Klarheit", Spektrum, Nr. 11 (1968), 373.

31. See e.g. E. Dahm, "Entwicklung und Festigung der neuen Beziehungen zwischen Arbeiterklasse und Intelligenz", Einheit, Nr. 10 (1959), 1354 ff.

32. C.W. Sherwin and R.S. Isenson, First Interim Report on Project Hindsight (Summary) (Washington, D.C.: Office of the Director of Defense Research and Engineering, 30 June 1966).

33. Traces. Technology in Retrospect and Critical Events in Science (Illinois Institute of Technology Research, Report Prepared for National Science Foundation, Washington, D.C., 1969).

34. J. Langrish, M. Gibbons, W.G. Evans and F.R. Jevons, Wealth from Knowledge. Studies of Innovation in Industry (London: Macmillan, 1972); M. Gibbons and R. Johnston, "The

Roles of Science in Technological Innovation", Research Policy, 3 (1974), 220-242; see also K. Pavitt, "Science, Enterprise and Innovation", Minerva, XI, No. 2 (1973), 275.

35. See Chapter Three and Appendix A.

36. Meske states that the Academy and higher education accounted for more than 10 % of the national economy's R&D manpower. W. Meske, "Zur Entwicklung des Forschungspotentials der DDR unter den Bedingungen der intensiv erweiterten Reproduktion im Sozialismus", in Sozialistische Wissenschaftspolitik und marxistisch-leninistische Wissenschaftstheorie, Kolloquien Reihe der Akademie, Heft 10 (1975), 121. On page 120 he gives the figure 12 %. Other sources give the Academy alone a proportion of 10 %. See: Klare and Hartkopf, op. cit., note 1, 25; "Das wissenschaftlich-technische Potential der DDR", Informatik, Nr. 3 (1973), 1.

37. Autorenkollektiv, Das Forschungspotential im Sozialismus (Berlin: Akademie Verlag, 1977), Table 2, 110.

38. International Survey of the Resources Devoted to R&D in 1969 by OECD Member Countries, Vol. 3, Private Non-Profit Sector (Paris: OECD, 1972), 54. Unfortunately, it is not possible to make comparisons for later years, as the statistics involved for West Germany group the graduates from engineering schools together with the QSEs. In the GDR such graduates would be classified as "Fachschulkader", which have been designated in this book as "qualified technicians". See ibid., 75; Survey of the Resources Devoted to R&D by OECD Member Countries. International Statistical Year 1971, Vol. 3, Private Non-Profit Sector (Paris: OECD, 1975), 73; Survey of the Resources Devoted to R&D by OECD Member Countries. International Statistical Year 1973, Vol. 3, Private Non-Profit Sector (Paris: OECD, 1977), 71.

39. See for instance Autorenkollektiv, op. cit., note 37. Table 3 on page 112 shows that in 1973, 46 % of the total employees of the Academy were "technicians". Table 4 on page 115 shows, however, that only 13.3 % of the total employees could be classified as "qualified technicians" ("Fachschulkader" and "Techniker").

40. H.-J. Böhme, "Unsere Aufgaben im Blick auf den IX. Parteitag der SED 1976", Das Hochschulwesen, Nr. 9 (1975), 274.

41. K. Hager, "Neue Etappe unserer Wissenschaftspolitik", Neues Deutschland (5. April 1969), 4.

42. L. Bode, "Die Einheit von Forschung und Lehre - ein wichtiges Prinzip zur Sicherung einer effektiven wissenschaftlichen Arbeit an Universitäten und Hochschulen", Wissenschaftliche Zeitschrift der Technischen Universität Dresden, 23, Heft 3/4 (1974), 526.

43. Survey of the Resources Devoted to R&D by OECD Member Countries. International Statistical Year 1973, Vol. 4, Higher Education Sector (Paris: OECD, 1977), 12 and 76.

44. Böhme, op. cit., note 40, 274 and 279.

45. H. Brennenstuhl and E. Schwertner, "Aufgaben der Forschung an den Universitäten und Hochschulen nach dem IX. Parteitag", Das Hochschulwesen, Nr. 10 (1976), 286.

46. G. Linnemann and S. Wykowski, "Erfahrungen und Zielstellungen im Zusammenwirken von Technischer Hochschule und Industrie auf dem Gebiet von Grundlagen- und angewandter Forschung", Das Hochschulwesen, Nr. 11 (1976), 306.
47. Autorenkollektiv, op. cit., note 37, 217.
48. Ibid., 215 and 45.
49. H.-J. Hicke, "Technika intensivieren Forschungsarbeit", Der Neuerer, Nr. 3 (1976), 98.
50. U. Hofmann, "Systematischer Ausbau der technisch-technologischen Basis", Spektrum, Nr. 10 (1975), 8.
51. N. Langhoff, "Wissenschaftlicher Gerätebau an der Akademie", Spektrum, Nr. 11 (1976), 26-28.
52. Hofmann, op. cit., note 50, 8.
53. U. Hofmann, "Wissenschaftspolitik und Forschungstechnologie", Spektrum, Nr. 6 (1976), 12.
54. Ibid.
55. Böhme, op. cit., note 40, 274.
56. H. Herpich, "Grundmittelanalyse an einer Universität", Sozialistische Finanzwirtschaft, Nr. 8 (1972), 48-49.
57. H. Seickert, J. Schindler and A. Selle, "Probleme der Überführung von Ergebnissen der Grundlagenforschung der Akademie der Wissenschaften der DDR in die materielle Produktion", Wirtschaftswissenschaft, Nr. 3 (1976), 422.
58. "Forschungsinstitute nutzen ihre Aggregate gemeinsam", Neues Deutschland (11. Februar 1976), 2.
59. P. Fiedler and G. Riege, "Das Anstrengendste steht noch bevor", Forum, Nr. 2 (1970), 9.
60. H. Klare, "Probleme und Maßnahmen für die weitere Entwicklung der Institute der Forschungsgemeinschaft", Spektrum, Nr. 4 (1962), 170 and 166.
61. H. Klare, Interview in Wissenschaft und Produktion, Nr. 5 (1969), 25.
62. G. Klose and M. Koss, "Durch sozialistische Forschungskooperation zu höherer Produktivität der wissenschaftlich-technischen Arbeit (Teil 1)", Die Technik, Nr. 5 (1972), 295.
63. See e.g. K. Pavitt and S. Wald, The Conditions for Success in Technological Innovation (Paris: OECD, 1971); and Gibbons and Johnston, op. cit., note 34.
64. See Chapter Three; also K. Eggert, "Hemmnisse in der Zusammenarbeit Universität - Industrie überwinden", Das Hochschulwesen, Nr. 7 (1964), 464.
65. St.J.DDR 1976, 62.
66. A. Meyer, "Kompromisse bei der Kombinatsbildung zahlen sich nicht aus", Die Wirtschaft, Nr. 28 (1970), 5.
67. H. Killiches, "Höhere Effektivität in Forschung, Technik, Produktion", Das Hochschulwesen, Nr. 7 (1972), 211.
68. W. Cziommer, "Formen und Inhalt der Zusammenarbeit von Hochschulen und Ingenieurschulen mit den volkseigenen Betrieben", Die Technik, Nr. 2 (1960), 59; Pavitt and Wald, op. cit., note 63, 94.
69. Klose and Koss, op. cit., note 62, 295. Geographical proximity is indicated by the names of the universities and the academy institutes.

70. H. Krieger, "Probleme des Kaderaustausches Hochschule - Praxis", Das Hochschulwesen, Nr. 4-5 (1970), 299.

71. H. Klare in Jahrbuch der Deutschen Akademie der Wissenschaften zu Berlin (1967), 147.

72. Krieger, op. cit., note 70, 296.

73. G. Caty, G. Drilhon, G. Ferné and S. Wald, The Research System. Volume I. France, Germany, United Kingdom (Paris: OECD, 1972), 230.

74. See for instance: "Direktorenkonferenz der Forschungsgemeinschaft", Spektrum, Nr. 6 (1964), 246.

75. J. Grabis, W. Liebig and J. Weiß, "Forschungsorganisation im Kombinat", Die Wirtschaft, Nr. 5 (1971), 13-14.

76. Cziommer, op. cit., note 68, 58; H. Klare, "Forschung nach Plan und Konzeption", Neues Deutschland (9. Februar 1964), 3.

77. W. Hartke, "Leibnitz-Tag, 30. Juni 1966", in Jahrbuch der Deutschen Akademie der Wissenschaften zu Berlin (1966), 87.

78. Die Deutsche Akademie der Wissenschaften auf dem Weg...", op. cit., note 17, 118-119.

79. Ebel, op. cit., note 30, 374.

80. Die Deutsche Akademie der Wissenschaften auf dem Weg...", op. cit., note 17, 25. NB: The number of institutes for natural science, technology and medicine was 71 in 1967 and 35 in 1975. See Hartkopf and Dunken, op. cit., note 1, 112-114; and Klare and Hartkopf, op. cit., note 1, 34-35.

81. See e.g. W. Bohn, "Wissenschaftsorganisatorische Probleme der Bildung eines Großforschungsverbandes", Das Hochschulwesen, Nr. 9-10 (1969), 643 ff.

82. Ebel, op. cit., note 30, 374; L. Gripinski, "Kooperationsbeziehungen zwischen Hochschule und Industrie", Das Hochschulwesen, Nr. 4 (1969), 249.

83. Die Deutsche Akademie der Wissenschaften auf dem Weg...", op. cit., note 17, 22.

84. Ebel, op. cit., note 30.

85. F. Liebscher, "Die Technische Universität Dresden auf dem Wege zur sozialistischen Universität", Die Technik, Nr. 10 (1969), 650.

86. Klare, Interview, op. cit., note 61, 23.

87. F. Hoche and H. Schäfer, "Beziehungen zwischen Hochschulen und Praxispartnern in der naturwissenschaftlich-technischen Forschung", Wirtschaftsrecht, 1, Nr. 1 (1973), 20.

88. See e.g. F. Selbmann, "Maßnahmen zur Verbesserung der Arbeit auf dem Gebiet der naturwissenschaftlich-technischen Forschung und Entwicklung und der Einführung der neuen Technik", Die Technik, Nr. 10 (1957), 667; H. Pöschel, "Forschung - Technik - Ideologie", Einheit, Nr. 3 (1966), 306.

89. W. Rühle, "Die Einbeziehung der Grundlagenforschung in den volkswirtschaftlichen Reproduktionsprozeß", Wirtschaftswissenschaft, Nr. 3 (1968), 354-355.

90. Gripinski, op. cit., note 82, 248.

91. E.A. Lauter, "Wissenschaftsorganisation an der Akdemie", Spektrum, Nr. 7 (1970), 7.

92. Linnemann and Wykowski, op. cit., note 46, 306.
93. Autorenkollektiv, op. cit., note 37, 156.
94. Survey of the Resources Devoted to R&D by OECD Member Countries. International Statistical Year 1971, Vol. 5, Total Tables (Paris: OECD, 1974), 31, 29, and 33.
95. One factor contributing to the relative lag of the GDR's chemical industry behind that of the FRG might be a substantially greater GDR dependence on external R&D. I am grateful to R. Amann for this point. Given the importance of the machine building industry for the GDR, the comparatively low proportion of contract finance from heavy machine and plant building is perhaps surprising but does not necessarily indicate particular technological backwardness. Institutionally, subjects relevant to machine building were by no means so well represented in the Academy as chemistry. Heavy machine and plant building was only part of the machine building sector, the other part being "processing machines and vehicle building". Moreover machine building was based in Karl-Marx-Stadt, Leipzig, Dresden and Magdeburg and it is probable that the industry had a fairly high level of contracts with the technical universities in these districts.
96. A. Scherzinger, Zur Planung, Organisation und Lenkung von Forschung und Entwicklung in der DDR (Berlin: Duncker und Humblot, 1977), 103-104.
97. Ibid., 104; Pohlisch, op. cit., note 14, 99. See comment on research "capacity" in note 15.
98. Of the OECD countries in 1971, Switzerland had the highest level of business enterprise supported university research at 8 %, and was followed by the UK at 5 %. Survey of the Resources..., op. cit., note 94.
99. Survey of the Resources Devoted to R&D by OECD Member Countries, International Statistical Year 1971, Vol. 5, Total Tables (Paris: OECD, 1975), 35.
100. Ibid., Vol. 1, Business Enterprise Sector, 104-110.
101. See Ruhle, op. cit., note 89, 359, 362-363; "Grundlagenforschung auch in der Industrie", Die Wirtschaft, Nr. 12 (1974), 7.
102. W. Kramer and W. Linden, "Forschungskooperation MLU-VVB Agrochemie", Das Hochschulwesen, Nr. 11 (1969), 772.
103. Klose and Koss, op. cit., note 62, 293.
104. Ibid., 294.
105. C. Grote, "Keine Pionier- und Spitzenleistungen ohne gesellschaftlichen Nutzen", Spektrum, Nr. 1 (1971), 22.
106. See e.g. Fiedler and Riege, op. cit., note 59, 9.
107. "Verordnung über die Leitung, Planung und Finanzierung der Forschung an der Akademie der Wissenschaften und an Universitäten und Hochschulen vom 23. August 1972", GBl, T II, Nr. 53 (1972), 592.
108. K. Hager, Sozialismus und wissenschaftlich-technische Revolution (Berlin: Dietz Verlag, 2. Auflage 1973), 17.
109. K. Hager, Wissenschaft und Technologie im Sozialismus (Berlin: Dietz Verlag, 1974), 20.

110. H. Klare, "Grundlagenforschung - Effektivität von übermorgen", Einheit, Nr. 10 (1972), 1283; also H. Klare, "Grundlagenforschung - Basis des wissenschaftlich-technischen Fortschritts", Einheit, Nr. 10 (1976), 1096-1101; and H. Klare and H. Müller, "Grundlagenforschung für Wissenschaft und Produktion", Wissenschaft und Fortschritt, Nr. 5 (1976), 196 ff.

111. "Ist unser wissenschaftlicher Vorlauf ausreichend?", Die Wirtschaft, Nr. 15 (1974), 4.

112. G. Prey, "Zur Entwicklung von Wissenschaft und Technik in der DDR bis 1975", Chemische Gesellschaft, 20, Nr. 1 (1973), 5.

113. Linnemann and Wykowski, op. cit., note 46, 305. See also: H. Seickert, "Fortschritte und Hemmnisse der Überführung", Spektrum, Nr. 10 (1975), 4; and H. Klare, "Wissenschaft im Auftrag und zum Nutzen unseres Volkes", Spektrum, Nr. 9 (1974), 10.

114. L. Kannengießer and W. Meske, "Wirksamer Einsatz des Forschungspotentials", Spektrum, Nr. 10 (1972), 19.

115. See for example: E. Schwertner, "Zur Intensivierung der wissenschaftlichen Arbeit an den Universitäten und Hochschulen", Das Hochschulwesen, Nr. 6 (1973), 164; H.-J. Böhme, "Die Aufgaben an den Universitäten und Hochschulen im Studienjahr 1976/77 in Auswertung des IX. Parteitages der Sozialistischen Einheitspartei Deutschlands", Das Hochschulwesen, Nr. 9 (1976), esp. 259-262; U. Hofmann and S. Heinrich, "Zur Zusammenarbeit zwischen der Akademie der Wissenschaften der DDR und dem Ministerium für Hoch- und Fachschulwesen", Das Hochschulwesen, Nr. 11 (1976), 307-311.

116. J. Müller, D. Schubert and B. Wilhelmi, "Überführung von Ergebnissen der physikalischen Forschung in die Produktion", Das Hochschulwesen, Nr. 10 (1974), 307. See a similar point for the Academy and universities in Klose and Koss, op. cit., note 62.

117. S. Wykowski, "Enge Verbindung von Grundlagenforschung und Überleitung der Forschungsergebnisse in die Praxis", Das Hochschulwesen, Nr. 4 (1976), 105.

118. K. Müller, "Phasen und Verkettungen der Überführung", Spektrum, Nr. 5 (1973), 17-18.

8 Conclusion

The German Democratic Republic is the most successful economy in Eastern Europe; it has the highest standard of living of these countries and the highest average level of technology. Its technology, particularly in machine building and chemicals, is of a respectable level in the world as a whole, although its technological performance has not matched that of its sister-state in the capitalist camp: the Federal Republic. Some reasons are suggested in the first section of this chapter for the GDR's technological lag behind the FRG, and in the second for the GDR's technological lead over the USSR. The final section contains some reflections towards a hypothetical model of technological change in the GDR.

THE GDR'S TECHNOLOGICAL LAG BEHIND THE FRG

The GDR appears to have had an especially pronounced lag behind the FRG in the more modern branches of technology such as electronics, data processing, instrument building, synthetic fibres and plastics. The lag is probably due more to endogenous than exogenous factors, and in this section some relevant systemic differences between the Germanies are summarised.

One set of differences concerns the research and development effort. Western analyses suggest that although Eastern Europe on the whole has lower levels of technology than the USA and Western Europe, its R&D inputs have been relatively high.[1] If GDR statistics for total and industrial R&D manpower were taken at face value, the GDR would seem to have had a smaller absolute but a larger relative R&D effort than the FRG. However GDR figures are considerably inflated when considered in terms of the OECD's Frascati definition. On a Frascati basis, the GDR's industrial R&D

effort in 1964 and 1970 was not only lower than that of the FRG in absolute terms, but also very likely lower in relative terms. Yet even if GDR figures were deflated by as much as 50 % to adjust for non-Frascati activities, the GDR would compare favourably with OECD countries other than the FRG. Data for 1964 indicate that the GDR placed a greater emphasis on machine building than the FRG and a significantly lower emphasis on chemicals. Too small a GDR research and development effort, at least up to the mid-sixties, in the "modern" branches of industry such as plastics, data processing, and electrical machine and apparatus construction was one major reason for their lower technological sophistication relative to West Germany. Another weakness of the GDR's effort relative to the FRG was the shortage of qualified technical staff. Further problems of the GDR's industrial and academic R&D effort seem to have been typical of the Soviet Union: the most highly qualified R&D personnel were located in the industrial institutes or the Academy of Sciences; there were difficulties of obtaining modern equipment and supplies; and the interface between development and production seems to have been rather weak.[2]

The second set of differences between the two Germanies concerns the influence of the political-economic system on the whole of the "science-production cycle", and especially on the introduction of new and improved technology into the enterprise. Although many of the obstacles to technological change evident in the GDR may also have played a retarding role in the FRG, it is unlikely that they were so pernicious there. The GDR was more insulated than its sister state from world developments in science and technology, and its system of technological change had weaker incentives and was more bureaucratic. These difficulties could only partly be compensated by pressures such as central directives, special priority plans, individual initiative, ideological emphasis of technological change, and the influence of mass organisations.

Such comments are not intended to read as an idealisation of the system of technological change in the Federal Republic. Observers in both East and West have noted that competition in the market can result in wasteful duplication of innovative activity, uneconomic non-specialisation of production, trivial product differentiation, and short-term R&D activity.[3] Wider economic, societal, psychological and moral problems of capitalist competition have, of course, been analysed and criticised by many social thinkers, and not least by Marx and Engels. Furthermore, over the years the West German government has become increasingly involved in promoting technological change but it is debatable as to how far this has been in a socially useful direction and undertaken in an economically effective manner. State expenditure on natural scientific-technical R&D increased in real terms by a factor of three between 1962 and 1972,[4] and each year it roughly matched the expenditure of the business enterprise sector.[5] According to a report published by the Ministry for Research and Technology in 1977, the three main research policy aims of the government were "modernisation of the economy", "improvement of living and working conditions", and "raising scientific efficiency, particularly in basic research".[6] Closer inspection of the statistics contained in the report, however, reveals that roughly 50 % of all governmental expenditure on R&D was allocated to "maintaining external security", nuclear power, space, data processing and aircraft.[7]

On average, the state financed an annual 14 % of the R&D conducted in the business enterprise sector over the eight-year period 1962-1969, and an annual 18 % over the succeeding eight-year period 1970-1977.[8] Most of this took the form of direct, i.e. project-based, support. One source estimated the ratio direct: indirect financial support for R&D (excluding finance from the Ministry of Defence) to have increased from 2.3 : 1 in 1970 to 16.2 : 1 in 1976.[9] The chief beneficiaries of direct governmental support were a handful of large firms. In 1976, 80 % of the governmental support from the Ministry for Research and Technology went to fifty large firms and 37 % to five of them. In 1973 the corresponding percentages were 93 % and 51 %.[10]

Some arguments in favour of this mode of governmental involvement are that only through direct support can the state purposefully influence technological change; that direct governmental support is necessary for international competitiveness; and that only large firms "are in the position to bear the high technical and economic risks of long-term R&D projects and to bring forth their own share of the necessary finance".[11] Some counter arguments are that firms are usually in a better position than governmental ministries to assess and reach decisions about new technology; that the privileged support of a few large research-intensive firms engenders important consequences for domestic competition; and that such support constitutes a socialisation of industrial risk and a diversion of state funds from economically and socially more useful areas. Moreover, if a large firm considered a project likely to be an economic and technical success, it would in all probability be able to find the financial resources necessary without government help.[12]

THE GDR'S TECHNOLOGICAL LEAD OVER THE USSR

With the exception of electrotechnology and electronics, civilian industry in the Soviet Union seems to have lagged behind that in the GDR by a substantial technological margin. In trying to account for the GDR's superior performance in technology, it is useful, as in the previous section, to consider factors relating firstly to the research and development effort and secondly to the political-economic system.

Research and development in the GDR shared some of the deficiencies of the Soviet system but also some of the advantages of the Western system. One similarity to the West was that most industrial R&D was based at or near the place of production. In the mid-sixties, the GDR had a fairly high proportion of industrial R&D manpower based in the enterprise, and the institutes which employed the remainder were usually situated near to the production units. By contrast, the majority of R&D personnel in the Soviet Union were located in institutes geographically separate from the factory.[13] Academic research establishments in the GDR were also geographically closer to industry than was the case in the USSR and probably specialised to a greater degree in subjects relevant to local industry. Another similarity to the West was that the administrative separation of industrial R&D and production in the GDR was not as great as that in the USSR.[14] From 1963 most of the institutes were subordinate to the VVBs (and later the combines) and not the ministries. The organisation of enterprises and institutes within the VVBs and combines was thus comparable in a formal sense to the organisation of a Western firm. In the seventies the Soviet Union made efforts to surmount the institutional barriers within R&D, and these efforts seem to have been in the direction of forming GDR-type VVBs and combines. Its new "ob'edinenie" grouped together enterprises, research institutes and design organisations, but differed widely with respect to the territorial scale of operation; relations between corporate management and constituent enterprises; range of output; branch of economy; the kinds of organisation included; and innovation objectives.[15]

Many of the GDR's policy measures for technological change have parallels, or appear to have been influenced by discussions in the USSR, and many of the hindrances to technological change seem to have been symptomatic of the Soviet system. Yet there were some important differences between the two countries. One was that the GDR inherited advanced skills after 1945. Pre-war Germany had been one of the strongest countries in research and technology, and these skills were not lost in the GDR despite the migration of a large number of qualified persons; indeed the GDR succeeded in building up a system of school and university education which has often been the envy of West Germany. Three other differences were the strength of economic signals, the level of decision making, and the pressure of competition.

Economic stimuli during the sixties were probably stronger in the GDR than the Soviet Union. The GDR introduced its New Economic System two years before the USSR and was in fact the first country in the Soviet bloc to embark upon major reform. Apart from Hungary, it was the country to go the farthest beyond the blueprint stage in implementing changes. "Benefit sharing" was applied in price formation in 1967, two years before the USSR; and flexible, degressive price construction was adopted four years before. The GDR introduced extended operative plan periods for bonuses in 1968, three years ahead of the Soviet Union, and its proportion of research financed by contracts was in all likelihood much higher.[16]

Decision making in the GDR after 1963 was more decentralised than in the USSR. It would seem that from the early sixties the GDR's political authorities were more enthusiastic than their Soviet counterparts about adopting new methods of economic organisation, and it is possible that the USSR regarded the GDR as something of an economic laboratory. The GDR's system of VVBs based on economic accounting methods and its subsequent system of combines antedate the Soviet "ob'edinenies" by ten and five years respectively. Whereas in the GDR a good deal of price-setting authority was devolved to the VVBs and sometimes the enterprises, in the USSR this authority remained fully in the hands of the State Price Committee.[17] The finance of technological change was also more decentralised in the GDR as a result of the "Fund for Technology" which was set up first at VVB level and later also at enterprise level. It is true that Soviet enterprises financed a certain amount of technological change themselves by including R&D charges against current costs of production; through their "Production Development Fund"; and by applying for State Bank credits. Most finance for technological change in the USSR, however, was centralised. The largest share of all R&D was supported by the state budget and the ministries; 80 % of all capital investment stemmed from or was earmarked by central agencies; and the "New Product Fund", although formed from enterprise contributions, was administered by the industrial ministries.[18]

Owing to the proximity of the FRG, the GDR has been subjected to a far greater degree of competition than has the USSR. The population of the GDR has always been aware of the standard of living in the other Germany, and this made legitimation of the GDR leadership a more acute problem than for the Soviet élite, perhaps rendering the former more efficiency conscious and more amenable to organisational change. Scientists, engineers and industrial managers were able to measure their achievements against, and learn about the latest results of, West German science and technology through FRG publications, the Leipzig Fair, and the import of FRG products.

None of the evidence collected for this study, however, indicates that the GDR moved in the direction of introducing competition between domestic R&D establishments. Some Western scholars noted that in the USSR complaints were made around the mid-sixties about the monopoly position of research institutes in certain industries.[19] Soviet proposals for competition were sanctioned in a 1968 decree which authorised ministries to assign R&D tasks to several organisations so as to pursue different approaches. The technique was to be employed however only "when necessary", for the "most important" scientific and technical problems. It applied only to the earliest stages of R&D work, and this work had to be "commissioned by the ministry".[20] In practice it may have been the case that GDR ministries sometimes worked along such lines for important projects and that enterprises sometimes exercised choice between different research institutes, but the explicit policy of the GDR was in the opposite direction: to concentrate R&D, to eliminate duplication, and to encourage cooperative links. "Socialist competition" is a term frequently used in the GDR, not least with respect to research and development, though in the first instance this principle amounts to the competition of groups or individuals against their plans. Where comparable work is being done by different groups such as in production, "socialist competition" can also involve the contest of one brigade or enterprise against another. In R&D the principle seems mainly to have involved the early completion of the Plan for Science and Technology, and the "overbidding", fulfilment or overfulfilment of scientific, technical and economic targets.[21]

Some interesting thoughts relating to technological change and the mode of enterprise management in the GDR and USSR have been expressed by the U.S. economist David Granick, and are the subject of the remainder of this section.[22] In his view, although the "orthodox, maximising" theoretical model of enterprise management seemed to have considerable explanatory value for Soviet industry, managerial behaviour in the GDR could be better described by the "satisficing model of American corporation-division behaviour". The "orthodox" model focused on bonuses for enterprise management and regarded managers as seeking to maximise their own bonuses in the long run by maximising specified quantitative objectives.[23] GDR practice, however, resembled the American in that managers appeared to "satisfice", i.e. 100 % fulfil, specific indicators and then use "their remaining resources to maximise aspects of performance not detailed in the plan targets set for them". Managerial bonuses were determined on subjective grounds in terms of the superior's judgement of overall performance rather than with respect to measured success.[24] Managerial careers in the GDR also resembled the American corporate pattern, and demotions were an "accepted part of the managerial milieu".[25]

Granick's prime source was interviews with persons holding managerial posts and he gives three main factors to account for the introduction of new products into production. One was that new models were introduced through the centralised planning of structure-determining tasks. Another was that producer enterprises were allowed to share in the economic usefulness of their new products. The third and "most important of all" was that enterprises were not faced with ambitious targets for specific indicators, nor did they attempt to overfulfil these targets. They therefore had reserves available for such "unmeasured-performance objectives" as the rapid introduction of new models.[26]

A criticism of the "orthodox model" which Granick does not make but which could be made is that it appears to ignore an important influence on management decision making: rewards for the workforce. It is not always the case that the bonus rules for management and workforce are congruent, as may be seen for the years 1968-1970 and 1971-1975 in the GDR. If, hypothetically, management bonuses rewarded the number of products of quality "Q" introduced into manufacture during the plan year and workforce bonuses rewarded quantitative production output, could one infer that the enterprise would attempt to implement a programme of rapid technological change? It is even possible that enterprise directors have been more concerned to maximise workforce bonuses than their own: indifference to rewards for enterprise personnel could have resulted in reprobation from the party organisation and trade union, pressure from the VVB or ministry, and even labour problems. Consequently the discussion of bonuses in this book dealt primarily with "enterprise bonuses", i.e. those for the workforce and for management.

Granick's alternative propositions on the mode of enterprise management in the GDR and the introduction of new technology are interesting and theoretically feasible but they receive little corroboration from the empirical data collected for this study. Firstly, the centralised planning of structure-determining tasks served to promote technological change, but more in the form of diffusion and imitation than the development and introduction of original GDR "models". Few such "models" would have appeared within the two years up to 1970 when Granick visited the GDR. Secondly, the procedure of benefit-sharing introduced in 1967 was an important improvement to the principles of price formation though, as noted in Chapter Five, it had a number of drawbacks in practice. Prices, at least prior to the order of May 1970, often did not contain a surcharge.

Thirdly, available evidence does not support Granick's most important proposition that during the NES, and particularly in 1970, technology was introduced as a combined result of central targets of low ambition; "satisficing" (i.e. 100 % fulfilment); reserves being used to fulfil "unmeasured" central objectives; and the threat of managerial demotion. Granted, until 1968 the national production plan does not seem to have been overly ambitious. Nevertheless the central planners tried to encourage enterprises not merely to "satisfice" main plan targets but to overfulfil them and indeed to adopt more ambitious ones than those prescribed. The bonus rules for these years show that overfulfilment of the main plan index resulted in a considerably increased bonus fund for the enterprise workforce. It also brought higher bonuses for enterprise management, since managerial performance was judged with particular reference to this index. In 1964 enterprises overfulfilling the profit target could direct up to 60 % of the overplan profit into their bonus funds; in 1965 and 1966 the percentage was up to 30 %.[27] A further device to encourage more than a mere "satisficing" of orientation indices was the "overbidding" procedure introduced in 1964. Detailed tables for calculating the size of the enterprise bonus fund according to the degree of overfulfilment and overbidding were published with the 1967 rules, and these show that the progressions were steep.[28] Between 1964 and 1968 it would seem that the central planners were interested in raising plan ambition, encouraging overfulfilment and reducing enterprise reserves. This interpretation is also consistent with the aggregate planned targets for the increase of industrial output. For 1964 the target was 5.4 %; for 1965, 5.7 %; for 1966, 5 %; for 1967, 6 %; and for 1968, 6.4 %.[29]

The bonus rules were different for the perspective periods 1969-1970 and 1971-1975. No planned profit target was set and hence no provisions made for "overfulfilment" and "overbidding". Enterprise workforce bonuses were based on the increase of absolute net profit together with the fulfilment of two side conditions. As there was no ratchet effect on profits within the perspective period, and enterprises were permitted to bring forward or carry over bonus funds from one year to another within the period, these rules aimed at stimulating bonus maximisation in the short run.

Enterprises undoubtedly set their own informal profit targets. It is therefore no surprise that in 1969, all enterprises of the machine building VVB Granick visited achieved at least 100 % fulfilment of their planned net profit.[30] However, non-fulfilment of the two indicators set as side conditions involved heavy deductions of the workforce bonus fund and available evidence strongly suggests that these indicators were ambitious in 1969 and 1970. Industrial output was planned to increase by 7 % in 1969 and 8 % in 1970, and there were many reports of supply problems and unfulfilled plans (including that for science and technology). Granick's own data accord with this argument. These data relate to three major enterprises of the machine building VVB and show that all three fulfilled their net profit targets but all suffered deductions from the maximum bonus fund earned.[31]

It is most unlikely that enterprises in 1970 had much in the way of reserves. Furthermore, the bonus law for that year supports neither the case that the introduction of new technology was an "unmeasured", "unquantified", or "unspecified" central objective,[32] nor that management bonuses were determined in a purely subjective fashion. Precisely some of the side conditions were intended to promote technological change, and the fulfilment of the conditions together with attention to contracts, exports and the continuity of planned output were to be used by the superior authorities to determine the level of management bonuses.[33] Granick might argue that the actual practice of awarding managerial bonuses differed substantially from that prescribed by the bonus law, but since no sign of this is apparent in the economic and enterprise literature the onus would then be upon him to provide supportive information relating to a larger number of enterprises than he in fact visited.

Granick states that GDR top managers seemed to be affected by career incentives more than in any other East European country. Demotion was "a serious threat": "I have data as to the next post of 15 predecessors of the top managers in my sample; 27 percent of them suffered clear demotion."[34] The corresponding percentages for the other countries he visited were Hungary, 17 %; Slovenia, 16 %; and Romania, 0.[35] Two points may be made about this conclusion. Firstly, the sample of "predecessors" was small. Secondly, the conclusion is critically dependent upon Granick's interpretation of industrial hierarchy in the various countries. He defined "positions at the same managerial level in a higher organisation" as "promotion" and "those at a substantially lower level there" as "demotion".[36] For the GDR he considered the combine to be a "higher organisation" than the enterprise, and the VVB "higher" than the combine. One can deduce from his figures that a total of four persons were "demoted", three of whom were VVB managers.[37] According to the above definitions, a person who transferred from a VVB to a similar managerial level in a combine was assumed to be "demoted". It is questionable, however, whether this assumption was justified. From 1968 the political leadership attached considerable importance to the creation of combines and the number of VVBs was reduced, though the reorganisations were more or less halted by the state of the economy in 1970. Data available for the year 1972 indicate that almost one-third of the combines were, like the VVBs, directly subordinate to the ministries. Moreover a new wave of combine formation took place between 1976 and 1979 leading to the disappearance of the VVBs. With such considerations in mind, and given that in 1970 there was by no means a surplus of highly qualified managerial personnel, it is not "clear" from Granick's data that managers in the GDR did face "a serious threat" of demotion.[38]

REFLECTIONS TOWARDS A MODEL OF TECHNOLOGICAL CHANGE IN THE GDR

Granick is right to question the "orthodox" theoretical Soviet model but his attempt to interpret GDR industrial management in terms of American corporate behaviour is not fully convincing, and his propositions do not satisfactorily account for the introduction of new technology into GDR industry. A contribution towards an analytical framework of technological change applicable both to socialist and capitalist economies has been made by another U.S. scholar, Joseph Berliner, in his study of the innovation decision in

Soviet industry. He focuses on the enterprise in an attempt to place it at a level coordinate with the enterprise in the capitalist economy, and proposes the general hypothesis that:

> Given the cultural and historical traditions, the technical characteristics, and the policies of government, the outcomes of any economy are fully explained in terms of four fundamental properties of economic structure: prices, decision rules, incentives and organisation.[39]

Accordingly, he states that one must look at these four properties to explain the rate of innovation in the USSR.[40]

An important point about Berliner's approach is that his hypothesis relates to only part of the process of technological change. His concept of "innovation" is given a fairly restricted meaning, referring to "those last links in the chain" of activities which begins with "a new idea in the mind of man" and concludes "with the attainment of the designed capacity of the new product or process".[41] A more comprehensive explanation of technological change would embrace earlier "links in the chain" (e.g. academic research) and include decision making at other levels of the economy besides the enterprise level. In particular, government policies could not be assumed to be "given". These remarks are pertinent whether the economy be of a socialist or capitalist type. Although Berliner holds that the hypothesis is applicable in both types, it would not satisfactorily explain technological change in such important areas as nuclear power, air transport and space.

Another point is that the four properties are probably not so independent as Berliner maintains.[42] Theoretically one property might be varied without affecting the other, though in practice this is unlikely. Owing to the assumed independence of these categories, however, and the fact that he presents his material under these headings, it is rather difficult for the reader to perceive the functioning of his system as a dynamic whole.

An alternative hypothesis might be based on the different interests of various groups in the economy. For a very rough understanding of the problems of technological change in the GDR, one could postulate a triangular system of relationships between the three broad interest groups: the party leadership, industry and the academic community. The first group was interested in technological change to achieve political legitimation; the second was primarily oriented towards production plan fulfilment; and the third was mainly interested in longer-term basic research. This resulted in a tension in the side of the triangle between the party leadership and industry. The former wanted technological change, but its understanding of the process was superficial, and its policies sometimes less than appropriate. The latter, although in a better position to promote and implement technological change, had little interest in doing so because of the lack of financial incentive. There was also tension in the second side of the triangle between the party leadership and academic community. Areas of contention included the respective levels of basic and applied research; individual, discipline-oriented research versus large-scale, problem-oriented research; internal self-government in academic science against external control by the party; and the extents of planning and contract work. The base of the triangle comprised a tension between the academic community and industry. This surfaced in a reluctance to cooperate, inadequate communication between the two sides, and complaints by each side about the other. In spite of measures taken to improve contacts, the important hindrance remained that industry did not have much incentive to pursue technological change actively.

A more satisfactory hypothesis would embrace a fourth broad interest group: the central authorities. It would note that the four groups were not discrete; the higher echelons of industry and of the academic community merged with the central authorities, and persons occupying top positions in the central agencies were also members of the party leadership. Such a hypothesis would also perceive the four groups as containing a plurality of interests and hence subgroups, these being variable over time particularly as a result of financial and organisational changes.

The academic community, for instance, consisted of the Academy group, which mainly conducted research, and of the university group, which was perhaps less prestigious and undertook both teaching and research. Both groups contained first and second generation academics; persons oriented towards the USSR and persons sympathetic to West Germany; "purists", who favoured the pursuit of pure science, "opportunists", who were interested in acquiring easy contracts; and yet others whose main function was to supervise the implementation of party policy.

Industry might be divided first into three horizontal levels: ministerial, VVB,[43] and enterprise. These levels could then be subdivided according to industrial branch, nature and size of undertaking, and type of department. The various interests within industry itself acted both to retard and to accelerate technological change.

Technological change could be retarded by ministerial interest in high output targets for its VVBs and enterprises, by the reluctance of one ministry or VVB to cooperate with another, and by the lack of financial motive for the VVB to impose enterprise profit cuts or price discounts. It was also retarded by the enterprise incentive system, by the interest of enterprise management in "stable" planning, and by the unwillingness of users either to criticise old products or to cooperate in the calculation of "benefit shares" for the producer. Further hindrances included the lack of enthusiasm of enterprise management both for R&D and the employment of graduates; the low esteem many people had for "technology"; the exclusion of users from "defence" proceedings; and the interest of industrial R&D establishments in pursuing topics of personal interest, topics which would assist the enterprise to fulfil its main success target, and topics containing "built-in" reserves.

Technological change was facilitated by each horizontal level of industry having an interest in concealing its true capabilities from the next higher level in an effort to secure "slack" output plans. It was also facilitated, given the sellers' market, by the interest of producer enterprises in inflating cost estimates. A number of groups in industry could exercise pressure via the planning mechanism on the enterprise to implement technological change. These included users; industrial ministries; VVB headquarters; and central or branch research institutes (or equivalent), the institutes being less oriented to the enterprise plan-year than the enterprise R&D establishments. Pressure outside of the planning mechanism may also have been applied on the enterprise through special directives and through individual initiative, especially when persons championed their R&D projects through to the manufacturing stage.

The central authorities contained several groups which could exercise a significant influence upon the rate of technological change. Besides the industrial ministries, important examples were the State Planning Commission, the Ministry for Science and Technology, the Research Council, the presidency of the Academy of Sciences, the Ministry of Higher Education, the Price Bureau, and the GBMQ. Over the years many changes have been made with regard to the demarcation of competence of these agencies and it is not possible to document and discuss the implications of these changes here. Some brief general remarks may, however, be made.

The State Planning Commission could promote technological change by prescribing plan indices for new or improved technology. It could retard technological change through taut production planning, use of the ratchet, and overcentralising planning responsibility. The latter resulted in a suffocatingly high level of paper work and at best only a statistical overview of research and technology topics.

The Ministry for Science and Technology helped to promote technological change by selecting projects of national importance and including them in the State Plan for Science and Technology, by monitoring the fulfilment of this plan, by preparing suitable documentation for top-level decision making, by exercising pressure on the industrial ministries, and by administering the Research Council whose function was to advise the State Planning Commission on questions of science and technology. The ministry may have retarded technological change by too little interest in basic research, although here counter pressure was exercised by the Academy presidency and the Ministry for Higher Education.

The Price Bureau facilitated technological change by awarding surcharges, by imposing discounts, and in general by being too overworked to check all price submissions. It could hinder the production of new technology by setting prices at too low a level and by making price revisions only infrequently. The GBMQ could stimulate technological change by checking product quality and initiating price adjustments. Unlike the enterprise or VVB it had no vested interest in demonstrating profitable production. Nevertheless, it could only deal with a limited number of products, and it retarded technological change by being too slow.

All of the central agencies mentioned above could positively influence the rate of technological change. Yet all shared the major disadvantage of being too distant from the enterprise to be fully aware of the types of technological change that were necessary or possible.

The most powerful interest group in the GDR was, of course, the party leadership. This group contained persons from the Politbureau, the Central Committee Secretariat, the State Council, and the Council of Ministers. It was responsible for policy formulation and for issuing important directives aimed at pushing through technological change. Policy decisions were functions of various power constellations within the group. Such constellations included the "pro and contra Ulbricht" factions, conservatives and liberals, and "planners and reformers". An obvious constraint to decision making was the necessity to secure the accordance, and sometimes obey the instructions of the Soviet Union: the USSR was responsible for the imposition of the detailed planning system in the GDR, gave the green light for the introduction of the New Economic System, and may also have had the last word on other policy measures.

Whereas many of the "planners" probably received their training in the Comintern schools of the USSR and in the fight against Fascism, the "reformers" were perhaps younger, trained in the GDR, and less tied down to a particular dogmatic position. It is tempting to speculate that the "offensive strategy" resulted in part from a top-level struggle between these two groups and in part from the psychological needs of Walter Ulbricht. The planners might have argued that the reformers had been given their chance with the NES, but this had failed to prove itself in practice and in any case had bordered on the ideologically suspect. Believing in the perfectibility of planning, the former group could have pointed to both the success achieved by the Soviet Union in the military and space sector through priority planning, and the extensive interest in the West in planning and other "rational" techniques such as operations research, computer methods, and scientific management. For Ulbricht the strategy perhaps represented a means of realising his ambition of presiding over the superior German state, of making the GDR a showcase of socialism, and of strengthening the country's position within Comecon: it held out the promise of overtaking West Germany in technology and, through its emphasis on central planning, provided a rostrum for criticising Czechoslovakian "revisionism".

The highly centralised political power at the disposal of the leadership meant there was a potential danger of naive, unsuitable or even arbitrary policy formulation. In some degree this danger was checked by two related factors. Firstly, owing to the volume of necessary and often complex decision making, the leadership was forced to depend on the various central agencies for advice and help. Secondly, although persons occupying the lower echelons of the industrial and academic hierarchies were usually denied a direct role in policy formulation, professionals and specialists could exercise indirect influence through publications and through comments passed upwards. The impact of these factors cannot be quantified and undoubtedly varied from one policy measure to another. There seems, for example, to have been a fair amount of discussion prior to the introduction of the New Economic System, and for many issues (such as whether prices should reflect use-value) members of the leadership would have been able to select from conflicting advice. The adoption of the "offensive strategy", on the other hand, suggests that the leadership could sometimes act unexpectedly, with little in the way of consultation and with little desire for subsequent critical observations or alternative proposals.

Published statements indicate that sections of the party leadership held rather uninformed views on technological change. In the early years of the GDR some representatives of the leadership believed that central planning would necessarily lead to a rapid rate of technological change. Both these and the top-level bureaucrats expected too great a degree of precision in the planning of research and technology, a common view being that all research results should be used; all new designs, introduced into production; and all plans for research and technology, fulfilled. In later years the leadership appeared to identify technological change with the generation of radical innovation. This they implicitly viewed as a one-directional linear process, having the sequence: basic research, applied research, development, and production. Although the leadership requested researchers to conduct work of importance to industry and the economy, their main emphasis was on scientific-technical opportunity rather than concrete user needs. Forecasting, which was heavily promoted in the sixties and especially between 1968 and 1971, was viewed as a new "link" at the beginning of this innovation chain.

There were a number of other consequences of the linear innovation conception. One was the call on academics in the seventies to demonstrate potential applications of their findings. Another was the interpretation of the "transfer phase" as being the weakest "link" in the "chain" when a substantial proportion of new applied research findings were not handed over into praxis. Such reasoning was put forward in favour of closer contacts between science and production, although important was not whether all research results were transferred but whether research was conducted or utilised to meet a specific need. A further consequence of the linear innovation idea was the demand for a "rational" and "efficient" system of organisation and management. In this respect one can observe some interesting parallels between research and industry. In both areas, similar methods of planning and financing were introduced. In both, there was allegedly too much "fragmentation". "Concentration" was called for, which in research meant larger establishments and research teams, fewer research topics and a division of labour between establishments. In industry, it meant larger enterprises, fewer products, and a division of labour between enterprises. The objective between 1968 and 1971 was to create "socialist large-scale research" and "socialist large-scale industry", the general assumption being that larger scale meant greater effectiveness.

Industrial innovation is admittedly a complex process. It would be reasonable to assert, however, that the party leadership's apparent conception of innovation was a superficial one. Every innovation in retrospect can probably be interpreted in terms of a linear chain of events, but closer scrutiny of an innovation's history would surely reveal a more complicated and less orderly situation, characterised by uncertainty, chance, time discontinuities and indeed "dialectic" interactions between various activities. Certainly the implicit conception neglected the importance of the user and misrepresented the role of basic research. It emphasised size at the expense of creativity and flexibility. In general it reflected a strong scientific determinism, which probably stemmed partly from the orthodox Marxist belief in science, and partly from the influence of Western advocates of "scientific management".

Theoretically, policy decisions of the leadership were supposed to be implemented according to the principle of "democratic centralism". This amounted, in effect, to the obedient fulfilment of the party line by the lower levels of the organisation concerned, with party activists, trade unions and other groups helping both to "transmit" policy decisions and supervise their implementation. An example of where this principle seems to have been applied resolutely in the GDR was the concentration of production. As Lenin noted:

> Large-scale mechanical industry - i.e. precisely the material, the productive source and basis of socialism - (requires the) unconditional and strictest unity of will ...[44] To govern, one needs an army of steeled revolutionaries, of communists. This army exists: its name is the Party.[45]

Often policy implementation took place less rigorously, particularly when measures were unclear, conflicting or impracticable. In the GDR, as in other countries, no policy could be formulated to cover all possibilities. This together with the frequently rushed preparation were important reasons for policies being unclear or imprecise. Consequently during the process of policy implementation, practitioners lower down in the hierarchy were to some extent obliged to engage in their own formulation of policy. The original policy could therefore undergo significant alteration.[46] This mechanism, however, served not only to check the implementation of inappropriate measures: it could also stifle useful ones. Thus inadequate delimitation of competence between ministry, VVB and enterprise led to inertia, buck-passing and arguments, and was probably an important reason why the efforts to devolve planning in the mid-fifties and again in the sixties were not very satisfactory.

In a similar fashion, policy underwent modification when measures were conflicting or impracticable. The enterprise was usually faced with a set of irreconcilable plan targets and since it could not fulfil all of them simultaneously it was forced to exercise choice. The VVB and ministry were in an analogous situation, and were therefore inclined to tolerate the fact that technological change generally took a subordinate place to production output. Impracticable measures were either ignored or paid lip-service. In the period 1968-1971, for instance, many researchers considered their results to be "pioneer achievements" and certain research establishments became "large-scale research centres" simply by a change of nameplate. Forecasts were often "knocked together" but not taken seriously, and the "structure-political conception" was probably based on surveys of development in the West rather than on any "highly qualified forecasts".

Given these problems, it is hardly surprising that the "transmission groups" were not always as "effective" as the leadership might have wished. In the sixties, especially, complaints were made about the lack of engagement of such groups in the promotion of technological change. Pöschel stated that some party secretaries measured the results of their work with scientists and engineers by the number of meetings or discussions held.[47] Berger complained that many trade unionists were one-sidely interested in the fulfilment of the production plan and neglected the role of research and development.[48] In the seventies, efforts were made to increase the influence of such groups in technological change, though essential conflicts of interest remained. The trade unions were supposed to execute SED policy on the one hand and to represent the interests of the workers on the other. There was also no clear line of demarcation between the responsibility of party management and that of the management of an R&D establishment or enterprise.

In short, the above hypothesis constitutes the beginnings of a model viewing technological change as a highly intricate process resulting from the different, variable and often conflicting forces of a complex, heterogeneous society. Perhaps its main advantage is that it serves to caution against oversimplified interpretations, whether from Eastern Europe in terms for example of the leading role of the party, broad social class, and the unity of interests in socialism; or from the West in terms of a few economic variables, totalitarianism, the party élite, naive marxist theory etc. Whilst pluralism undoubtedly exists in the GDR, it differs in both scope and mode to that of West Germany: some well-known examples of the limits to GDR pluralism have been demonstrated by the cases of Benary, Behrens, Harich, Havemann, Biermann and Bahro.

Such a hypothesis is also a useful frame of reference when considering potential obstacles to reforms aimed at significantly accelerating the rate of technological change in the GDR. Conceivably, at least three far-reaching reforms would be necessary. One would involve a substantial decentralisation of decision-making competence to lower levels of the economy; the actual level depending upon factors such as the particular decision involved, the branch of industry, and the likely economic and social consequences for other sections of the national economy. Another reform would introduce the possibility of freely exercising criticism at each level of society. A third would concern a much greater use of the market mechanism, and the restriction of central planning to the main and longer-term macroeconomic quantities. These reforms would entail practicable solutions being found to some profoundly difficult economic problems, not least the introduction of competition, the effective synchronisation of market and plan, and the avoidance of inflation, unemployment and duplicate investments. The greatest obstacle, however, would surely be the opposition from various vested interests.

Upper-level bureaucracies would scarcely welcome measures which drastically curtailed their power empires, and cut down their staff. Enterprise and combine management might be apprehensive about the increased responsibility of working within a market system. Party dogmatists at all levels, fearing for their positions, would object to ideologically suspect measures, and even some of the more progressive party members would be sincerely concerned that a new economic mechanism might jeopardise the achievements of the GDR, particularly with respect to full employment. Local party functionaries would feel demoralised if their sphere of activity, especially in management and the organisation of supplies outside the planning system, were reduced in favour of more authority for the enterprise and combine management, and a greater role for prices and contracts. In all probability such interest groups would avoid a stance of outright opposition: instead they would seek to undermine the reforms by a strategy of quiet obstinacy and reluctant cooperation.

The Soviet leadership would certainly not give any unqualified support to reform proposals. On the one hand the USSR has a material interest in the GDR raising its level of technology, and an ideological one in the GDR improving its standard of living. On the other hand contemporary history demonstrates that the Soviet Union would tolerate economic reform only in so far as it did not threaten to spill over into political instability.

Similar considerations apply for the GDR leadership. Too slow a rate of economic growth, and stagnant or falling living standards could induce unrest in the population; an effective stimulation of technological change through a socialist market economy with democratic decision-making processes could lead to political collapse. Given these constraints one might expect future policies of the GDR leadership to be directed mainly towards organisational and planning improvements[49] and perhaps a modest use of "economic levers" along the lines of the New Economic System. Prospects of a consequent disappearance or even narrowing of the technology gap between the two Germanies are therefore slim.

> Socialism has an unlimited capacity for perpetual renewal; it can continually adapt its concrete societal relationships to changing requirements, including those of scientific-technical progress. In this sense there are no barriers to scientific-technical progress stemming from the relationships of society in socialism.
> By contrast, the societal, social problems which the scientific-technical revolution intensifies under the relationships of capitalism can only be solved by the overthrow of this societal order.[50]

NOTES

1. J. Slama and H. Vogel, "Comparative Analysis of Research and Innovation Processes in East and West", in C.T. Saunders (ed.), Industrial Policies and Technology Transfer between East and West (Wien and New York: Springer, 1977), 106 f. The first systematic attempt to compare the R&D efforts of the USA and USSR was made by Freeman, Young and the Birmingham group for the year 1962. A revised and extended version was later completed by Davies and Berry of Birmingham and this indicated that the Soviet Union had probably been devoting a substantially larger manpower effort to R&D than the USA for a number of years up to and including 1966. On the other hand, the Birmingham investigation of Soviet industry published in 1977 showed that over the preceding 15-20 years, there had been no evidence of a substantial diminution of the technological gap between the USSR and the West. Similar conclusions were also reached by Kosta, Kramer and Slama in 1971 with regard to Czechoslovakia and Austria. Their comparisons suggested a considerably higher R&D effort in terms of manpower and expenditure for the former country, but higher technological levels for the latter. See: C. Freeman and A. Young, The Research and Development Effort in Western Europe, North America and the Soviet Union (Paris: OECD, 1965); R.W. Davies and M.J. Berry, "The Research and Development Effort in the Soviet Union: A Reconsideration of the Manpower Data", in Science Policy in the USSR (Paris: OECD, 1969), 501-557; R. Amann, J. Cooper and R.W. Davies (eds.), The Technological Level of Soviet Industry (New Haven and London: Yale University Press, 1977); H.G.J. Kosta, H. Kramer and J. Slama, Der technologische Fortschritt in Österreich und in der Tschechoslowakei (Wien: Springer Verlag, 1971). NB: Owing to the lack of published Soviet and Eastern European statistics, most East-West comparisons of R&D to date have been on an aggregate basis: industrial and especially branch R&D could not be separated out. Fortunately some disaggregate data were available for the GDR.

2. R. Amann, J. Berry and R.W. Davies, "Science and Industry in the USSR", in E. Zaleski et al., Science Policy in the USSR (Paris: OECD, 1969), 378-557.

3. See e.g. R.W. Davies, "Research, Development and Innovation in the Soviet Economy, 1968-1970", in D.O. Edge and J.N. Wolfe (eds.), Meaning and Control. Social Aspects of Science and Technology (London: Tavistock, 1973), 244.

4. H. Echterhoff-Severitt, Forschung und Entwicklung in der Wirtschaft 1971 (Essen: Stifterverband für die deutsche Wissenschaft, 1974), Table 3, 33.

5. Bundesminister für Forschung und Technologie, Faktenbericht 1977 zum Bundesbericht Forschung (Bonn: Bundesminister für Forschung und Technologie, 1977), Table 2, 215.

6. Ibid., 33.

7. In 1977 the figure amounted to 51.4 % of total state R&D expenditure. The figure planned for 1980 amounted to 48.5 %. See statistics in Faktenbericht 1977, op. cit., note 5, 36-64.

8. Calculated from Faktenbericht 1977, op. cit., note 5, Table 3, 216-217.

9. "Innovation. Ideen zum Erfolg", Wirtschaftswoche, Nr. 17 (21. April 1978), 58.

10. Faktenbericht 1977, op. cit., note 5, 15.

11. Ibid., 14.

12. K. Pavitt, "Government Policies Towards Innovation: A Review of Empirical Findings", Omega, No. 5 (1976), 539-558; and K. Pavitt and W. Walker, "Government Policies Towards Industrial Innovation: A Review", Research Policy, 5 (1976), 11-97.

13. Amann, Berry and Davies, op. cit., note 2.

14. Ibid.

15. J. Berliner, The Innovation Decision in Soviet Industry (Cambridge, Mass.: The MIT Press, 1976), 131-132.

16. R. Amann, "La Recherche Soviétique des Anées 70", La Recherche, Nr. 29 (1972), 1032 ff.; Davies, op. cit., note 3, 241-261; Berliner, op. cit., note 15. NB: Berliner notes that "the bulk of the income of R&D organisations continues to come from direct budget grants rather than from contracts", p. 116.

17. Berliner, op. cit., note 15, 390.

18. Ibid., Chapter Six.

19. Amann, Berry and Davies, op. cit., note 2, 454 f.; Berliner, op. cit., note 15, 120-129.

20. Berliner, op. cit., note 15, 126.

21. See e.g.: Autorenkollektiv, Die Ökonomie der betrieblichen Forschung und Entwicklung (Berlin: Verlag Die Wirtschaft, 1976), 211-212; R. Dittrich and J. Steiner, "Erfahrungen bei der Organisation des sozialistischen Wettbewerbs in produktionsvorbereitenden Bereichen", Sozialistische Arbeitswissenschaft, Nr. 5 (1973), 345-352; H. Leger and I. Waltenberg, "Wissenschaftlich-technische Arbeit im Blickfeld der Parteikontrolle", Einheit, Nr. 11 (1975), 1304-1308; H. Lysk, "Zielstrebige politisch-ideologische Arbeit beschleunigt wissenschaftlich-technischen Fortschritt", Einheit, Nr. 7 (1975), 714-718.

22. a) D. Granick, Enterprise Guidance in Eastern Europe. A Comparison of Four Socialist Economies (Princeton, N.J.: Princeton University Press, 1975); b) D. Granick, "Variations in Management of the Industrial Enterprise in Socialist Eastern Europe", in Reorientation and Commercial Relations of the Economies of Eastern Europe, A Compendium of Papers submitted to the Joint Economic Committee Congress of the United States (Washington, D.C.: U.S. Government Printing Office, 1974), 229-247.

23. See also Berliner, op. cit., note 15, 403-404.

24. Granick, op. cit., note 22a, 481.

25. Ibid., 483.

26. Ibid., 482.

27. "Beschluß über die Bildung und Verwendung des einheitlichen Prämienfonds in den volkseigenen und ihnen gleichgestellten Betrieben der Industrie und des Bauwesens und in den VVB im Jahre 1964", GBl, T II, Nr. 10 (1964), 83-84; "Beschluß über die Grundsätze für die Bildung und Verwendung des einheitlichen Prämien-

fonds in der volkseigenen Wirtschaft im Jahre 1965", GBl, T II, Nr. 42 (1965), 297-299; "Beschluß zur Richtlinie für die Bildung und Verwendung des Prämienfonds in den volkseigenen und ihnen gleichgestellten Betrieben und den VVB der Industrie und des Bauwesens im Jahre 1967 sowie zur Übergangsregelung für das Jahr 1966", GBl, T II, Nr. 40 (1966), 253.

28. GBl, T II, Nr. 40 (1966), 254-256. NB: One could, however, speak of a "satisficing" of "overbid" indices. The enterprise did not receive a larger bonus fund for "overfulfilling" an "overbid" index.

29. M. Keren, "The New Economic System in the GDR: An Obituary", Soviet Studies, Nr. 4 (1972/73), 578.

30. Granick, op. cit., note 22a, 199.

31. Ibid., 199-200.

32. Ibid., 483.

33. "Verordnung über die Bildung und Verwendung des Prämienfonds in den volkseigenen und ihnen gleichgestellten Betrieben, volkseigenen Kombinaten, den VVB (Zentrale) und Einrichtungen für die Jahre 1969 und 1970", GBl, T II, Nr. 67 (1968), 492.

34. Granick, op. cit., note 22b, 236.

35. Granick, op. cit., note 22a, 452.

36. Ibid., 452.

37. Ibid.

38. The problem of organisational definitions occurs elsewhere in Granick's account, in particular with regard to his comparison of the degree of central administration in the GDR and Romania. He estimates that 0.51 % of all industrial workers and employees in the GDR were occupied with "the coordination of the activities of operating units" and this was twice the proportion found in Romania. "This is not what we would expect in a system devoted to decentralising authority to the Kombinat and enterprise level" (note 22a, p. 162). It is also surprising when compared with Granick's conclusions that the Romanian system was "exceedingly centralised" (p. 477) and that in the GDR "there is considerable decentralisation of authority and responsibility to the enterprise and Kombinat level" (p. 478). For the GDR, Granick took the "coordination personnel" to consist of persons in the ministries, the VVBs and the product groups in combines and enterprises. For Romania, this personnel was taken to comprise only of persons in the ministries.

39. Berliner, op. cit., note 15, 18.

40. Ibid., 20.

41. Ibid., 2-3.

42. Ibid., 11-13.

43. As in Chapter Five, the combine is largely excluded here in the interests of clarity. In the first wave of combine building from 1968 to 1971, combines could be subordinate either to the ministry or the VVB. In the second wave from 1976 to 1979, the VVBs were abolished.

44. W.I. Lenin, "Die nächsten Aufgaben der Sowjetmacht", in Werke, Band 27 (Berlin: Dietz Verlag, 1972), 259.

45. W.I. Lenin, "Gesamtrussischer Verbandstag der Bergarbeiter", in Werke, Band 32 (Berlin: Dietz Verlag, 1961), 48.

46. For a good discussion of this mechanism with reference to the U.S. context see: C.E. Lindblom, The Policy-Making Process (Englewood Cliffs, N.J.: Prentice-Hall, second edition 1980).

47. H. Pöschel, Leitung von Forschung und Technik und wissenschaftlicher Meinungsstreit (Berlin: Dietz Verlag, 1964), 57.

48. R. Berger, "Mit dem sozialistischen Wettbewerb unsere Zukunft sichern", Die Arbeit, Nr. 1-2 (1965), 13.

49. One might interpret the replacement of VVBs by combines from this perspective.

50. H. Nick, Wissenschaftlich-technische Revolution (Berlin: Dietz Verlag, 1975), 44.

Appendix A

Tables

Table A.1
Manpower in Scientific-Technological R&D

	1963/64	1965	1968	1970	1971	1973	1974	1975
Total	100,000[a]	92,000[b]	90,000[d]	155,386[e]	155,000[f]	155,000[g] 150,000[h]	155,000[i]	155,000[j] 150,000[k]
of which QSE + QT		39,000[c]	33,000[d]		78,000[c,f]	80,000[h]	80,000[i]	78,000[j] 75,000[k]

Sources:

a) A. Springer, "Die Effektivität der wissenschaftlichen Forschung erhöhen!", Einheit, Nr. 6 (1966), 818; H. Pöschel, Leitung von Forschung und Technik und wissenschaftlicher Meinungsstreit (Berlin: Dietz Verlag, 1964), 51; H. Such, VVB und wissenschaftlich-technischer Fortschritt (Leipzig: Staatsverlag der Deutschen Demokratischen Republik, 1964), 127.

b) In 1965, 1.2 % of all employees were engaged in natural scientific-technological R&D. See W. Meske, "Zur Entwicklung des Forschungspotentials der DDR unter den Bedingungen der intensiv erweiterten Reproduktion im Sozialismus", Sozialistische Wissenschaftspolitik und marxistisch-leninistische Wissenschaftstheorie, Kolloquien Reihe der Akademie der Wissenschaften der DDR, Institut für Wissenschaftstheorie und -organisation, Heft 10 (1975), 116. The number of employees in 1965 was 7,675,800, see St.J.DDR 1970, 57.

c) Informatik, Nr. 4 (1972), 9; "Entwicklung des wissenschaftlich-technischen Potentials der DDR", Die Wirtschaft, Nr. 7 (1972), 20.

d) D. Graichen and L. Rouscik, Zur sozialistischen Wirtschaftsorganisation (Berlin: Verlag Die Wirtschaft, 1971), 225.

e) In 1970, 2 % of all employees were engaged in R&D. See Meske, op. cit., note b, 120-121. There were 7,769,300 employees in 1970. St.J.DDR 1972, 57.

f) A. Lange, "Den Zyklus 'Wissenschaft - Technik - Produktion' beherrschen (Teil I)", *Die Technik*, Nr. 1 (1973), 3.

g) "Wissenschaft und Technik entscheiden wesentlich über das Wachstum der Volkswirtschaft", *Statistische Praxis*, Nr. 10 (1973), 498.

h) "Das wissenschaftlich-technische Potential der DDR", *Informatik*, Nr. 3 (1973), 1; Autorenkollektiv, *Die Intensivierung der sozialistischen Industrieproduktion und die wachsende Rolle der Arbeiterklasse* (Berlin: Dietz Verlag, 1975), 58.

i) *Science in the Service of the People* (Berlin: Panorama DDR - Auslandspresseagentur, 1974), 58.

j) A. Lange and D. Voigtberger, *Überleitung von wissenschaftlich-technischen Ergebnissen* (Berlin: Verlag Die Wirtschaft, 1975), 9.

k) K.U. Brossmann and A. Lange, "Forschung und Entwicklung mit Investitionen und Grundfondsreproduktion einheitlich planen", *Die Technik*, Nr. 10 (1975), 660.

Table A.2
Formal Comparison of Total R&D Manpower
(excluding Social Sciences and Humanities) in GDR and FRG

	GDR	FRG	GDR as % of FRG
1963/64	90,000 - 100,000[a]	186,901[c]	48.1 - 53.5
1973	150,000 - 155,000[b]	303,837[d]	49.4 - 51.0
% increase	50 - 72.2	62.6	

Sources:

a + b) Table A.1.
c) A Study of Resources Devoted to R&D in OECD Member Countries in 1963/64, Vol. 2, Statistical Tables and Notes (Paris: OECD, 1968), 48.
d) Survey of the Resources Devoted to R&D by OECD Member Countries, International Statistical Year 1973, Vol. 5, Total Tables, Statistical Tables and Notes (Paris: OECD, 1976), 106.

Table A.3
Formal Comparison of Industrial R&D Manpower in GDR and FRG

	GDR	FRG	GDR as % of FRG
1963/64	65,100[a]	127,928[c]	50.9
1969		169,990[d]	
1970	> 85,146[b]		> 50.1
% increase	> 30.8	32.9	

Sources:

a) First section of Chapter Three.
b) In 1970 more than 3 % of industrial employees were engaged in R&D. See W. Meske, "Zur Entwicklung des Forschungspotentials der DDR unter den Bedingungen der intensiv erweiterten Reproduktion im Sozialismus", in Sozialistische Wissenschaftspolitik und marxistisch-leninistische Wissenschaftstheorie, Kolloquien Reihe des Instituts für Wissenschaftstheorie und -organisation an der Akademie der Wissenschaften der DDR, Heft 10 (1975), 121. The number of industrial employees in 1970 was 2,838,200, St.J.DDR 1971, 53. NB: This figure is slightly different to that quoted in later statistical yearbooks owing to adjustment of definitions (c.f. St.J.DDR 1976, 47 and 53). The difference, however, is unimportant for this comparison.
c) A Study of Resources Devoted to R&D in OECD Member Countries in 1963/64, Vol. 2, Statistical Tables and Notes (Paris: OECD, 1968), 158.
d) International Survey of the Resources Devoted to R&D in 1969 by OECD Member Countries, Statistical Tables and Notes, Vol. I, Business Enterprise Sector (Paris: OECD, 1972), 82. NB: FRG figures represent "manufacturing" with "mining".

Table A.4
Industrial R&D Manpower in 1963/64

Absolute Figures		Relative Figures (Per 1,000,000 Population)	
Japan	171,487	FRG	2,221
FRG	127,928	Sweden	2,203
France	69,753	Japan	1,788
Italy	21,804	France	1,458
Sweden	16,750	Belgium	1,269
Canada	13,274	Netherlands	799
Belgium	11,791	Canada	701
Netherlands	9,567	Austria	554
Austria	3,970	Italy	432
GDR (unadjusted)	65,100	GDR (unadjusted)	3,795

Sources:

A Study of Resources Devoted to R&D in OECD Member Countries in 1963/64, Vol. 2, Statistical Tables and Notes (Paris: OECD, 1968), 158-159. The OECD items "mining" and "manufacturing" have been grouped together to approximate the GDR definition of "industry". Figures for Switzerland, the UK and USA are not available.
Absolute GDR figure from Table A.3.
Population figures from St.J.DDR 1965, 14*-15*. The population figure for the FRG includes West Berlin.

Table A.5
Ratio of Industrial R&D Employees to Total R&D Manpower in 1971 (%)

GDR	> 54.9
Belgium	62.4
Canada	33.3
Finland	49.8
France	53.7
FRG	69.1
Greece	18.6
Italy	53.2
Japan	60.3
Norway	44.1
Spain	31.6
Sweden	57.7

Sources:

Calculated from Survey of the Resources Devoted to R&D by OECD Member Countries, International Statistical Year 1971, Vol. 1, Business Enterprise Sector (Paris: OECD, 1975), 111, and Vol. 5, Total Tables (Paris: OECD, 1974), 64. The items "mining" and "manufacturing" have been grouped together to approximate the GDR definition of "industry". NB: Figures for Switzerland, the UK and USA are not available.
GDR statistic from Table A.1 and Table A.3.

Table A.6
Distribution of Persons Employed in R&D Establishments Over the Sections of the National Economic Council (1 January 1963)

	Total Employed (FTEs) in %
Energy	0.7
Coal	1.3
Iron and Steel Metallurgy	1.3
Non-Iron and Steel Metallurgy and Potash	1.9
Foundry and Forging	0.8
State Geological Commission	0.5
Chemicals	13.7
Heavy Machine Building	18.4
General Machine Building	13.4
Electrotechnical/Electronics	22.7
Machine Tools and Automation	14.1
R&D Establishments directly subordinate to Machine Building	5.1
Textiles, Clothing, Leather	1.2
Wood, Paper, Printing	0.9
Glass and Ceramics	2.3
Food	1.6
Others	0.1
Total	100.0

Source: H. Kusicka and W. Leupold, _Industrieforschung und Ökonomie_ (Berlin: Dietz Verlag, 1966), 39.

Note: The breakdown corresponds to a definition of industry employed by the National Economic Council, and is not identical to that used by the Statistical Yearbook for production statistics. See e.g. _St.J.DDR 1964_, 84.

Table A.7
Industrial R&D Manpower in Chemicals 1963/64

Absolute Figures		Relative Figures (Per 1,000,000 Population)	
Japan	48,155	FRG	777
FRG	44,739	Belgium	510
France	13,787	Japan	502
Italy	8,224	France	288
Belgium	4,740	Netherlands	246
Canada	2,950	Sweden	229
Netherlands	2,943	Italy	163
Sweden	1,745	Canada	156
Austria	929	Austria	129
GDR (unadjusted)	8,919	GDR (unadjusted)	520

Sources:

A Study of Resources Devoted to R&D in OECD Member Countries in 1963/64, Vol. 2, Statistical Tables and Notes (Paris: OECD, 1968), 158-159. The OECD items "chemicals", "petroleum refining and extraction", "drugs" and "rubber products" have been grouped together to approximate the GDR definition of "chemical industry" (c.f. breakdown in St.J.DDR 1964, 92).
Absolute GDR figure is based on Table A.6 and Table A.3.
Population figures from St.J.DDR 1965, 14*-15*. The population figure for the FRG includes West Berlin.

Table A.8
Ratio of R&D Manpower to Work Force in Several VVBs
(1 January 1964)

VVB	%
Synthetic Fibres and Photochemicals	1.94
Electrochemicals and Plastics	2.11
General Chemicals	2.70
Mineral Oils and Organic Basic Materials	2.79
Rubber and Asbestos	1.70
Pharmaceutical Industry	3.78
Paints and Varnish	2.44
Plastics Processing	0.98
Mining Equipment and Conveyance Plant	4.40
Chemical Plant	4.66
Textile Machine Building	2.73
Polygraph, Machines for Paper & Printing	3.68
Data Processing and Office Machines	2.73
Food and Packing Machines	3.75
High Tension Equipment and Cable	3.42
Components and Vacuum Technology	3.86
Machine Tools	3.54
Automatic Control Equipment Construction & Optics (excluding VEB Carl Zeiß)	4.96
Communications and Measurement Technology	6.80
Electrical Machine Construction	3.24
Electrical Apparatus Construction	2.50
VEB Carl Zeiß Jena	6.70

Source: H. Kusicka and W. Leupold, Industrieforschung und Ökonomie (Berlin: Dietz Verlag, 1966), 41.

Table A.9
Proportion of University Graduates in Total R&D Manpower (%)
(Selection of VVBs and VVB Institutes in 1963/64)

VVB High Tension Equipment and Cable	15.4
VVB Components and Vacuum Technology	13.3
VVB Polygraph, Machines for Paper and Printing	11.1
VVB Data Processing and Office Machines	6.9
Institute for Electronics	29.7
Institute for Chemical Plant	21.4
Institute for Printing Machines	20.0
Institute for Machine Tools	14.9
Institute for Conveying Technology	8.4
Institute for Automatic Control Technology	7.8

Source: H. Kusicka and W. Leupold, Industrieforschung und Ökonomie (Berlin: Dietz Verlag, 1966), 47.

Table A.10
Estimated Distribution of Industrial R&D Manpower in 1972 (%)

Electrotechnical, Electronics & Instrument Building	35.6
Machine & Vehicle Building	26.0
Chemicals	24.0
Light Industry, Textiles & Food	7.8
Basic Materials excluding Chemicals	6.7

Source: Estimate based on the diagram contained in G. Kehrer, "Wechselbeziehungen zwischen territorialer Produktions- und Wirtschaftsstruktur", in H. Richter (ed.), Beiträge zur territorialen Produktionsstruktur (Gotha/Leipzig: VEB Hermann Haack, Geographisch-Kartographische Anstalt, 1976), 56.

Note: To minimise errors in measurement the diagram was photographically enlarged by a factor four.

Table A.11
R&D Personnel in Firms 1964 to 1971 — FRG*

Firms with ... to ... Employees	Year	Employees in R&D			
		Total		of which Scientific Personnel and R&D Management	
		Number	%	Number	%
under 500	1964	4,381	5.1	576	5.2
	1965	3,482	3.7	443	3.7
	1967	3,534	3.9	425	3.4
	1969	2,405	2.0	289	1.7
	1971	3,489	2.5	394	2.1
500 - 1,999	1964	8,258	9.6	1,002	9.1
	1965	8,533	9.2	955	7.9
	1967	9,584	10.5	1,254	10.0
	1969	8,437	7.2	1,169	7.1
	1971	12,632	9.0	1,842	9.9
2,000 and over	1964	73,051	85.3	9,472	85.7
	1965	80,948	87.1	10,643	88.4
	1967	78,051	85.6	10,827	86.6
	1969	106,753	90.8	15,127	91.2
	1971	123,838	88.5	16,443	88.0

Source: H. Echterhoff-Severitt, Forschung und Entwicklung in der Wirtschaft 1971 (Essen: Stiferverband für die deutsche Wissenschaft, 1974), 19. Reprinted by Permission.

*Excluding firms in the chemical industry.

Table A.12
Employees in the German Academy of Sciences 1949-1980

Year	Number	Source
1949	< 1,000	a
1950	1,000	b
1952	2,110	c
1955	4,000	d
1956	5,270	c
1957	6,302	c
1960	> 8,000	a, b
1961	10,000	d
1965	12,000	e
1970	> 13,000	a
1974	> 15,000	a
1975	> 16,000	b, f
1980	19,000 (planned)	g

Sources:

a) W. Meske, "Zur Entwicklung des Forschungspotentials der DDR", in *Sozialistische Wissenschaftspolitik und marxistisch-leninistische Wissenschaftstheorie*, Kolloquien Reihe des Instituts für Wissenschaftstheorie und -organisation an der Akademie der Wissenschaften der DDR, Heft 10 (1975), 109.
b) Autorenkollektiv, *Das Forschungspotential im Sozialismus* (Berlin: Akademie Verlag, 1977), 115.
c) Quoted in R. Landrock, *Die Deutsche Akademie der Wissenschaften zu Berlin 1945-1971*, Band 1 (Erlangen: Deutsche Gesellschaft für zeitgeschichtliche Fragen e.V.), 83.
d) H. Klare and W. Hartkopf, *Die Akademie der Wissenschaften der DDR. Zum 275. Jahrestag der Gründung der Akademie* (Berlin, 1975), 21.
e) W. Hartke, "Leibniz-Tag", in *Jahrbuch der Deutschen Akademie der Wissenschaften zu Berlin* (1966), 86.
f) H. Laitko and G. Wendel, "Wissenschaftspolitische Wandlungen", XV. Internationaler Kongreß für Geschichte der Wissenschaft, Edinburgh, August 1977, 13.
g) U. Hofmann, "Wissenschaftspolitik und Forschungstechnologie", *Spektrum*, Nr. 6 (1976), 10.

Table A.13
Students Graduating from University in the GDR and FRG by Subject of Study 1962-1967/69 (%)

	GDR			FRG	
	1962	1967	1969	1962	1967
Maths., Natural Sciences[a]	6.2	7.4	6.5	5.2	5.5
Technical Sciences	14.9	20.4	20.2	10.5	9.3
Medicine, Agriculture[b]	19.8	20.8	19.0	13.1	15.0
Economics[c]	15.8	7.9	12.8	7.9	10.2
Philosophy and History,[d] Politics and Law	5.1	1.8	3.0	10.5	8.9
Culture, Aesthetics, Sport, Literature and Languages	3.0	2.7	3.2	5.4	5.4
Art	1.9	1.8	1.5	4.2	3.9
Secondary School Teaching[e]	33.4	37.0	33.7	43.4	41.8
	100	100	100	100	100

Source: From Tabelle A124 in: Bundesministerium für innerdeutsche Beziehungen, *Bericht der Bundesregierung und Materialien zur Lage der Nation 1971* (Kassel, 1971), 414-415.

Notes:

a) Figures for FRG do not include psychologists.
b) Figures for FRG do not include food chemists.
c) Figures for FRG include student teachers for schools of commerce.
d) Figures for FRG include educationalists ("Pädagogen") and psychologists.
e) Figures for GDR include all basic subjects of education.

Table A.14
Industrial Contracts to Academic Research in the GDR and FRG in 1971 (%)

GDR		FRG		
Industrial Contracts to Academy of Sciences		Industrial Contracts to Universities and Research Institutes of Bund and Länder		
Chemicals	44.2	Steel/Machine/Vehicle Construction		63.5
Electrotechnology/ Electronics	28.1	of which Aircraft	29.6	
Heavy Machine and Plant Building	8.5	" " Machine Building	29.4	
		Chemicals		13.3
		Transport/Communications		6.8
		Electrotechnology/Fine Mechanics/Optics		5.6
Total R&D Funds from Industrial Contracts	100	Total R&D Funds from Industrial Contracts		100

Sources:

Calculated from Autorenkollektiv, Das Forschungspotential im Sozialismus (Berlin: Akademie Verlag, 1977), 156; H. Echterhoff-Severitt, Forschung und Entwicklung in der Wirtschaft 1971 (Essen: Stifterverband für die deutsche Wissenschaft, 1974), 62-63.

Note:

Figures for the FRG are based on columns 6 and 8 of Echterhoff-Severitt. A rough approximation to the FRG's definition of "industry" is obtained by including "construction" in the GDR's definition and then normalising the figures. For an explanation of the FRG definition see Echterhoff-Severitt, p. 128.

Appendix B

Experimental Comparison of Industrial R&D in the GDR and FRG for 1964 and 1970 According to the Frascati Definition

METHOD

The first step involves the exclusion of persons from the GDR statistics who would not be classified as R&D manpower in the "Frascati" sense. It is very probable that GDR statistics are inflated by persons providing an indirect service to R&D. In 1964, for instance, "administrative personnel, maintenance staff, enterprise guard, drivers, cleaning staff and others" were included in the statistics.[1] Many of these workers would not come within the Frascati definition of R&D manpower. Although a rough figure for the size of this category is available for 1964, none could be found for 1970. This problem can, however, be avoided to a large extent if one makes the ratio of qualified scientists, engineers and technicians (QSE and QT) to total industrial employees the basis of measurement, rather than the ratio of R&D manpower to total industrial employees.

The second step involves the calculation of the number of qualified scientists, engineers and technicians in the GDR for the years 1964 and 1970. According to Table 3.5 in the text, these workers accounted in 1964 for 34.4 % of total industrial R&D manpower. Since there were about 65,100 persons performing industrial R&D in 1964, the figure for QSE and QT is 22,394. In 1970 "more than 3 % of industrial employees" were said to be engaged in R&D. It is extremely unlikely that this proportion was more than 3.4 % as otherwise the author would have cited a higher figure.[2] The number of industrial R&D workers comes to between 85,146 and 96,499. An estimate must be made of the number of (QSE + QT)s in 1970. It is noted here that in 1965 the ratio (QSE + QT) : (R&D manpower) was roughly the same in industry as it was for the national economy, i.e. 34 % compared with 42 % (see Table A.1). In the early seventies the ratio for the national economy was about 50 % (Table A.1). It may therefore be assumed that the ratio for industry in 1970 lay between 34 % and 50 %. The overall lower and upper limits for QSE and QT in 1970 thus come to 28,950 and 48,250.

In FRG statistical yearbooks, figures for "industrial employees" include "apprentices", whereas in GDR statistical yearbooks the items "industrial employees" and "apprentices" are listed separately. The third step therefore is to calculate the "total industrial employees" in the GDR for the years 1964 and 1970.

The fourth step involves the adjustment of West German statistics on R&D manpower (OECD) and industrial employees (Statistical Yearbooks) to correspond roughly with the GDR definition of "industry": this is done by grouping together the categories "manufacturing" and "mining". It is noted, however, that the West German R&D statistics possibly underestimate the real situation since the surveys were not official and depended upon the goodwill of the respondents.[3]

The fifth step concerns the adjustment of GDR R&D statistics to full-time equivalents. As noted in Chapter Three, the research institutes engaged not only in R&D but also production work, standardisation, information and documentation. The R&D performed by these institutes also involved a certain amount of routine work such as obtaining supplies. Estimates of the amount of time spent on routine matters by R&D workers in such institutes varied between 20 % and 50 % for the period 1964-1966. Enterprise R&D establishments also engaged in non-R&D activities, particularly production: in many enterprises in 1966 between 60 % and 70 % of "design and technological capacity" was used for current production. It would be unwise to overinterpret these estimates but they do indicate that a fair proportion of activity was not R&D in the Frascati sense. Such figures were not found in the literature for the early seventies though it is clear from the criticisms published that a similar situation existed. Nor are figures available for the early seventies on the size of R&D manpower by type of establishment (i.e. institute or enterprise). The problem of reducing GDR R&D statistics to full-time equivalents is dealt with here by postulating an overall lower limit for the proportion of routine/non-Frascati activity in the statistics. It would be reasonable to assume that the total non-Frascati activity accounted for no less than one-quarter of the total working time of industrial R&D manpower in the years 1964 and 1970, and in all probability was considerably higher.

Observers of R&D in the Soviet Union have also been faced with the problem of trying to assess the extent of routine or non-Frascati work included in the official statistics. In the Freeman and Young study of the R&D effort in Western Europe, North America and the Soviet Union published in 1965, Davies, Barker and Fakiolas assumed that all the manpower of the industrial research institutes were conducting R&D in a Frascati sense.[4] For the so-called "project and design organisations" two estimates were made. In one, 50 % of their manpower was conducting R&D and in the other estimate, none. The 1969 study by Davies and Berry stated that more detailed examination of Soviet R&D organisations had shown, amongst other things, that industrial R&D institutes were to some extent also engaged in non-R&D activities.[5] This corresponds to the situation in the GDR as outlined above. The authors, however, could not give statistics on the proportions of non-Frascati work, and had to resort to estimates. For the research institutes, a low estimate of 10 % routine work was suggested, and a high estimate of 50 %. For the project-design establishments, 30 % was suggested. These figures, taken together with proportions for other types of establishment, were later rounded off to upper and lower limits of 40 % and 20 % respectively.

The final stage of this comparison of the two Germanies involves the calculation of that multiplier which adjusts GDR industrial research intensity to the FRG level. If the multiplier is higher than 0.75 then it is reasonable to conclude that the GDR R&D effort was lower in a relative sense than that of the FRG, both since we have assumed a minimum of 25 % non-Frascati work and since the FRG figures are probably somewhat too low.

Detailed statistics for this calculation are given below, and the results are that the multiplier had the value 1.12 in 1964, and 0.75 to 1.24 in 1970. Thus, both in 1964 and 1970 the GDR very likely had a relatively lower industrial R&D effort than the FRG.

CALCULATION

As explained above, the basis of the comparison is the equation:

$$\frac{(\text{QSE} + \text{QT})_{\text{GDR}}}{(\text{Total Industrial Employees})_{\text{GDR}}} \times F = \frac{(\text{QSE} + \text{QT})_{\text{FRG}}}{(\text{Total Industrial Employees})_{\text{FRG}}}$$

If the multiplier "F" is greater than 0.75, it is reasonable to conclude that the GDR not only had a smaller absolute industrial R&D effort than the FRG, but also a smaller relative industrial R&D effort.

Figures for the GDR

Table B.1
Qualified Scientists, Engineers and Technicians in the GDR

	1964	1970
Total No. in Industrial R&D	65,100[a]	85,146[c] - 96,499[d]
QSE + QT	22,394[b]	28,950[e] - 48,250[f]

Notes:

a) See Chapter Three.
b) 34.4 % of total industrial R&D was QSE + QT. See Table 3.5.
c) Assuming 3 % of industrial employees were engaged in industrial R&D. The number of industrial employees in 1970 was 2838.2×10^3. St.J.DDR 1972, 57.
d) Assuming 3.4 % of industrial employees were engaged in industrial R&D.
e) For an overall lower limit assume 34 % of industrial R&D manpower was QSE + QT.
f) For an overall upper limit assume 50 % of industrial R&D manpower was QSE + QT.

Table B.2
Total Industrial Employees in the GDR (including apprentices)

	1964	1970
Employees (excl. apprentices)	2791.5×10^3 [a]	2838.2×10^3 [b]
Apprentices	144.9×10^3 [c]	194.5×10^3 [b]
Total Employees	2936.4×10^3	3032.7×10^3

Sources:

a) St.J.DDR 1970, 57.
b) St.J.DDR 1972, 57.
c) St.J.DDR 1966, 63.

NB: In 1968/69 the GDR adopted a somewhat different classification of industry. With the exception of the figure for apprentices in 1964, all figures in the table are according to the post-1968 classification system. No figure for the number of apprentices in 1964 according to the new system could be found so the old one was used.

Figures for the FRG

Table B.3
Qualified Scientists, Engineers and Technicians in the FRG

	1964	1969
QSE in Manufacturing	16,097	47,245
QSE in Mining	328	523
QT in Manufacturing	53,804	50,366
QT in Mining	674	578
QSE + QT	70,903	98,712

Sources:

A Study of Resources Devoted to R&D in OECD Member Countries in 1963/64, Vol. 2, Statistical Tables and Notes (Paris: OECD, 1968), 162 and 164; *International Survey of the Resources Devoted to R&D in 1969 by OECD Member Countries*, Vol. 1, Business Enterprise Sector (Paris: OECD, 1972), 86 and 88.

Table B.4
Total Industrial Employees in the FRG (including apprentices)

	1964	1969
Employees in Manufacturing (incl. apprentices)	7804.5×10^3	7991×10^3
Employees in Mining (incl. apprentices)	496.7×10^3	318×10^3
Total Employees	8301.2×10^3	8309×10^3

Sources: *St.J.BRD 1966*, 228; *St.J.BRD 1970*, 187.

Using these data to calculate the multiplier for 1964 and 1969/70 we have:

1964

$$\frac{22,394}{2936.4 \times 10^3} \times F = \frac{70,903}{8301.2 \times 10^3}$$

$$F = 1.12$$

1969/70

$$\frac{28,950}{3032.7 \times 10^3} \times F = \frac{98,712}{8309 \times 10^3} \qquad \frac{48,250}{3032.7 \times 10^3} \times F = \frac{98,712}{8309 \times 10^3}$$

$$F = 1.24 \qquad\qquad F = 0.75$$

From these results it would seem most likely that the GDR had a relatively lower industrial R&D effort than the FRG both in 1964 and in 1970.

NOTES

1. See Chapter Three.
2. W. Meske, "Zur Entwicklung des Forschungspotentials der DDR unter der Bedingung der intensiv erweiterten Reproduktion im Sozialismus", in Sozialistische Wissenschaftspolitik und marxistisch-leninistische Wissenschaftstheorie, Kolloquien Reihe des Instituts für Wissenschaftstheorie und -organisation an der Akademie der Wissenschaften der DDR, Heft 10 (1974), 121.
3. OECD, A Survey of Resources Devoted to R&D in OECD Member Countries in 1963/64, Vol. 2, Statistical Tables and Notes (Paris: OECD, 1968), 18; OECD, International Survey of the Resources Devoted to R&D in 1969 by OECD Member Countries, Vol. 1, Business Enterprise Sector (Paris: OECD, 1972), 130-132.
4. C. Freeman and A. Young, The Research and Development Effort in Western Europe, North America and the Soviet Union (Paris: OECD, 1965), 28. R.W. Davies, G.R. Barker and R. Fakiolas were responsible for the appendix on the Soviet Union.
5. R.W. Davies and M.J. Berry, "The Research and Development Effort in the Soviet Union: A Reconsideration of the Manpower Data", in Science Policy in the USSR (Paris: OECD, 1969), 508.

List of Abbreviations

A	Austria
AEG	Allgemeine Elektricitäts-Gesellschaft (Electrical Firm in FRG)
B/L	Belgium and Luxemburg
Comecon	Council for Mutual Economic Assistance
DDR	Deutsche Demokratische Republik (See GDR)
DIW	Deutsches Institut für Wirtschaftsforschung (German Institute for Economic Research)
DK	Denmark
DM	Deutsche Mark
EEC	European Economic Community
FuE	Forschung und Entwicklung (See R&D)
FRG	Federal Republic of Germany
GB	Great Britain
GBl	Gesetzblatt (Law Gazette)
GBMQ	German Bureau for Measurements and Quality Control
GDR	German Democratic Republic
GNP	Gross National Product
I	Italy
M	Mark
MIT	Massachusetts Institute of Technology
N	Norway
NES	New Economic System
n.e.s.	not elsewhere specified
NL	Netherlands
OECD	Organisation for Economic Cooperation and Development
QSE	Qualified Scientist and Engineer
QT	Qualified Technician
R&D	Research and Development
SED	Sozialistische Einheitspartei Deutschlands (Socialist Unity Party of Germany)
SITC	Standard International Trade Classification
STC	Scientific-Technical Centre
St.J.BRD	Statistisches Jahrbuch der Bundesrepublik Deutschland (Statistical Yearbook of the FRG)

St.J.DDR	Statistisches Jahrbuch der Deutschen Demokratischen Republik (Statistical Yearbook of the GDR)
T	Teil (Part)
TOM	Technical-Organisational Measure
UK	United Kingdom
UNESCO	United Nations Educational, Scientific, and Cultural Organisation
USA	United States of America
USSR	Union of Soviet Socialist Republics
VEB	Volkseigener Betrieb (State-Owned Enterprise)
VVB	Vereinigung Volkseigener Betriebe (Association of State Enterprises)

Bibliography

ARTICLES, BOOKS, LAWS, REPORTS AND STATISTICS

A bis Z. Ein Taschen- und Nachschlagebuch über den anderen Teil Deutschlands (Bonn: Deutscher Bundesverlag, 1969).

Akademie der Wissenschaften der DDR. Zentralinstitut für Geschichte, DDR - Werden und Wachsen (Berlin: Dietz Verlag, 2., durchgesehene Auflage, 1975).

"Akademie, Hochschulen und Industrie gründen Technika", Neues Deutschland (17. Dezember 1975), 4.

"Akademie-Industrie-Komplexe gegründet", Spektrum, Nr. 7 (1976), 33.

Amann, R., "The Soviet Research and Development System: The Pressures of Academic Tradition and Rapid Industrialisation", Minerva, 8, No. 2 (1970), 217-241.

Amann, R., "La Recherche Soviétique des Anées 70", La Recherche, Nr. 29 (1972), 1027-1034.

Amann, R. and J. Slama, "The Organic Chemicals Industry of the USSR: A Case-Study in the Measurement of Comparative Technological Sophistication by Means of Kilogram-Prices", Research Policy, 5 (1976), 302-326.

Amann, R., J. Berry and R.W. Davies, "Science and Industry in the USSR", in E. Zaleski et al., Science Policy in the USSR (Paris: OECD, 1969), 378-557.

Amann, R., J. Cooper and R.W. Davies (eds.), The Technological Level of Soviet Industry (New Haven and London: Yale University Press, 1977).

Amann, R. and J. Cooper (eds.), Industrial Innovation in the Soviet Union (New Haven and London: Yale University Press, 1982).

Ammer, T., "Reform der Deutschen Akademie der Wissenschaften zu Berlin", Deutschland Archiv, Nr. 5 (1970), 546-551.

Anders, W., "10 Jahre Zentralinstitut für Schweißtechnik der DDR", Die Technik, Nr. 5 (1962), 353-357.

"Anordnungen" see under "Gesetzblätter".

"Antworten aus der Praxis", Die Wirtschaft, Nr. 33 (1968), 8-10.

"Die Anwendung der fortgeschrittensten Wissenschaft und die Herstellung der Rentabilität in der Industrie", Neues Deutschland, (B), (19.4.1955), 3.

Apel, E., "Schlußwort zur Diskussion über das Referat und die vorgelegten Entwürfe der Dokumente", in W. Ulbricht, Das Neue Ökonomische System der Planung und Leitung der Volkswirtschaft in der Praxis (Berlin: Dietz Verlag, 1963), 261-290.

Apel, E. and G. Mittag, Wissenschaftliche Führungstätigkeit - Neue Rolle der VVB (Berlin: Dietz Verlag, 1964).

Ardenne, M. von, Wege zur Steigerung der Weltmarktfähigkeit unserer industriellen Erzeugnisse (Berlin: Verlag Die Wirtschaft, 1963).

Arnold, H., H. Borchert, A. Lange and J. Schmidt, Die wissenschaftlich-technische Revolution in der Industrie der DDR (Berlin: Verlag Die Wirtschaft, 1967).

"Auch bei Konsumgütern Planerfüllung sichern", Die Wirtschaft, Nr. 15 (9. April 1970), 2.

Die Aufgaben der Universitäten und Hochschulen im einheitlichen Bildungssystem der sozialistischen Gesellschaft. IV. Hochschulkonferenz, 2. und 3. Februar 1967 in Berlin (Berlin: Staatsverlag der Deutschen Demokratischen Republik, 1967).

Die Aufgaben zur weiteren ökonomischen Stärkung der DDR und zur Festigung der sozialistischen Demokratie (Berlin: Dietz Verlag, 1961).

"Aus dem Bericht des Politbüros an die 14. Tagung des ZK der SED", Neues Deutschland (10. Dezember 1970), 4-8.

"Aus dem Bericht über die Ergebnisse der Untersuchung zu Problemen der Weiterentwicklung der Planung in den Betrieben und Kombinaten im ökonomischen System des Sozialismus", Die Wirtschaft, Beilage zur Ausgabe Nr. 18 (1968), 22.

"Aus den Materialien der 10. Tagung des ZK der SED", Die Wirtschaft, Beilage zur Ausgabe Nr. 20 vom 15. Mai 1969, 2-12.

Autorenkollektiv, Die Finanzen der Industrie in der Deutschen Demokratischen Republik (Berlin: Verlag Die Wirtschaft, 1966).

Autorenkollektiv, Ökonomische Probleme des wissenschaftlich-technischen Fortschritts (Berlin: Verlag Die Wirtschaft, 1966).

Autorenkollektiv, Neuererbewegung - Arbeiterinitiative zur sozialistischen Rationalisierung (Berlin: Staatsverlag der Deutschen Demokratischen Republik, 1973).

Autorenkollektiv, Sozialistische Betriebswirtschaft. Lehrbuch (Berlin: Verlag Die Wirtschaft, 2. Auflage 1974).

Autorenkollektiv, Die Intensivierung der sozialistischen Industrieproduktion und die wachsende Rolle der Arbeiterklasse (Berlin: Dietz Verlag, 1975).

Autorenkollektiv, Die Ökonomie der betrieblichen Forschung und Entwicklung (Berlin: Verlag Die Wirtschaft, 1976).

Autorenkollektiv, Das Forschungspotential im Sozialismus (Berlin: Akademie Verlag, 1977).

Banse, T. and H. Nick, "Gebrauchswert und Preisbildung", Wirtschaftswissenschaft, Nr. 6 (1975), 843-850.

Bartsch, H. and G. Kraft, "Prognostik ist kein Hilfsprozeß", Die Wirtschaft, Nr. 47 (1970), 13-14.

Bauerfeld, H. and R. Weiss, Aktuelle Probleme der Neuerer-

bewegung (Berlin: Verlag Tribüne, 1972).
Behrendt, K.H., "Der ökonomische Nutzeffekt bei der Bildung volkseigener Kombinate", Sozialistische Finanzwirtschaft, Nr. 12 (1970), 11-15; and Nr. 13 (1970), 27-30.
Behrendt, K.H. and W. Heidel, "Leistungsbewertung und wissenschaftlich-technischer Fortschritt", Sozialistische Finanzwirtschaft, Nr. 11 (1975), 12-15.
Berger, R., "Mit dem sozialistischen Wettbewerb unsere Zukunft sichern", Die Arbeit, Nr. 1-2 (1965), 12-17.
Bergsdorf, W., "Produktionsfaktor Wissenschaft. Zur Wissenschafts- und Forschungspolitik der DDR", Die neue Gesellschaft, Nr. 2 (1968), 167-175.
"Bericht über eine Beratung zu Problemen der Elektrotechnik in der Deutschen Demokratischen Republik", Einheit, Nr. 11 (1959), 1498-1509.
"Bericht des Zentralkomitees an den VI. Parteitag der sozialistischen Einheitspartei Deutschlands", Neues Deutschland (11.10.1962), 3-10.
Berliner, J., The Innovation Decision in Soviet Industry (Cambridge, Mass.: The MIT Press, 1976).
Bernicke, E., "Technologie und wissenschaftlich-technische Revolution", Einheit, Nr. 9 (1967), 1107-1115.
Berry, M.J., "Soviet Research and Development after the 1965 Reforms", Jahrbuch der Wirtschaft Osteuropas, Band I (1970), 252-264.
"Beschluß des Ministerrates über Rolle, Aufgaben und die weitere Entwicklung der Deutschen Akademie der Wissenschaften zu Berlin vom 27. Juni 1963", Spektrum, Nr. 11/12 (1963), 374-377.
"Beschluß des Staatsrates der Deutschen Demokratischen Republik über weitere Maßnahmen zur Gestaltung des ökonomischen Systems des Sozialismus", Die Wirtschaft, Beilage zur Ausgabe Nr. 18 (1968), 4-6.
"Beschlüsse" see under "Gesetzblätter".
Betriebsökonomik, 3. Teil: Industriekaufleute (Berlin: Verlag Die Wirtschaft, 1965).
Betriebsökonomik Industrie. Teil 2 (Berlin: Verlag Die Wirtschaft, 7., überarbeitete Auflage 1971).
Block, G., "Tendenzen in der Forschungsorganisation der Industrie", Die Technik, Nr. 5 (1969), 299-303.
Blume, P., "Der Plan Neue Technik als Voraussetzung zur planmäßigen Erfüllung des Rekonstruktionsplanes", Die Technik, Nr. 10 (1961), 673-681.
Bode, L., "Die Einheit von Forschung und Lehre - ein wichtiges Prinzip zur Sicherung einer effektiven wissenschaftlichen Arbeit an Universitäten und Hochschulen", Wissenschaftliche Zeitschrift der Technischen Universität Dresden, 23, Heft 3/4 (1974), 525-530.
Böhme, H.-J., "Zum Stand der Erfüllung der Beschlüsse des VIII. Parteitages an den Universitäten und Hochschulen", Das Hochschulwesen, Nr. 9 (1973), 262-269.
Böhme, H.-J., "Unsere Aufgaben im Blick auf den IX. Parteitag der SED 1976", Das Hochschulwesen, Nr. 9 (1975), 258-280.

Böhme, H.-J., "Die Aufgaben an den Universitäten und Hochschulen im Studienjahr 1976/77 in Auswertung des IX. Parteitages der Sozialistischen Einheitspartei Deutschlands", Das Hochschulwesen, Nr. 9 (1976), 248-268.

Bohn, W., "Wissenschaftsorganisatorische Probleme der Bildung eines Großforschungsverbandes", Das Hochschulwesen, Nr. 9-10 (1969), 643-648.

Borchert, H., "Zu einigen Fragen der Planung 'Neue Technik' im VEB Elektrochemischen Kombinat Bitterfeld", Wissenschaftliche Zeitschrift der Martin-Luther-Universität Halle-Wittenberg, Ges.-Sprachwissenschaftliche Reihe, XI, Nr. 7 (1962), 817-822.

Borchert, H., "Die Finanzierung des wissenschaftlich-technischen Fortschritts und die ökonomischen Hebel seiner schnellen Durchsetzung", Wissenschaftliche Zeitschrift der Martin-Luther-Universität Halle-Wittenberg, Sonderheft 1964, 9-29.

Böttcher, M., "Wo stehen wir mit der Planung im NÖS?", Die Wirtschaft, Nr. 48 (1965), 16-17.

Böttcher, M., "Die optimale Betriebsgröße - Scheinproblem oder Mittel ökonomischer Diagnose?", Die Wirtschaft, Nr. 48 (1966), 10-12.

Brennenstuhl, H. and E. Schwertner, "Aufgaben der Forschung an den Universitäten und Hochschulen nach dem IX. Parteitag", Das Hochschulwesen, Nr. 10 (1976), 284-287.

Breshnew, L.I., Rechenschaftsbericht des Zentralkomitees der KPdSU an den XXIV. Parteitag der KPdSU (Moskau: APN-Verlag; Berlin: Dietz Verlag, 1971).

Brossmann, K.U. and A. Lange, "Forschung und Entwicklung mit Investitionen und Grundfondsreproduktion einheitlich planen", Die Technik, Nr. 10 (1975), 659-663.

Buch, G., Namen und Daten (Berlin and Bonn - Bad Godesberg: J.H.W. Dietz Nachf. GmbH, 1973).

Bundesminister für Forschung und Technologie, Faktenbericht 1977 zum Bundesbericht Forschung (Bonn: Bundesminister für Forschung und Technologie, 1977).

Bundesminister für innerdeutsche Beziehungen, "Die Entwicklung des innerdeutschen Handels", Information, Nr. 5 (1979), 12-13.

Bundesministerium für innerdeutsche Beziehungen, Bericht der Bundesregierung und Materialien zur Lage der Nation 1971 (Kassel, 1971).

Bundesminister für wissenschaftliche Forschung, Bundesbericht Forschung II (Bonn: Bundesminister für wissenschaftliche Forschung, 1967).

Bunge, M., "Kybernetik - Operationsforschung - Netzwerktechnik - EDV", Die Wirtschaft, Nr. 33 (1968), 19.

Buzmann, E., "Forschungsarbeit und Praxisverbindungen aus der Sicht der Ingenieurhochschule", Das Hochschulwesen, Nr. 8 (1974), 249-251.

Caty, G., G. Drilhon, G. Fernê and S. Wald, The Research System. Volume I. France, Germany, United Kingdom (Paris: OECD, 1972).

Cooper, J., "Research, Development and Innovation in the Soviet

Union", in Z.J. Fallenbuchl (ed.), Economic Development in the Soviet Union and Eastern Europe, Volume 1 (New York: Praeger, 1975), 159-195.

Cooper, J., "The Scientific and Technical Revolution in the USSR", Co-existence, No. 2 (1981), 175-192.

Cziommer, W., "Formen und Inhalt der Zusammenarbeit von Hochschulen und Ingenieurschulen mit den volkseigenen Betrieben", Die Technik, Nr. 2 (1960), 57-61.

Dahm, E., "Entwicklung und Festigung der neuen Beziehungen zwischen Arbeiterklasse und Intelligenz", Einheit, Nr. 10 (1959), 1348-1362.

Davies, R.W., "Science and the Soviet Economy." An Inaugural Lecture delivered in the University of Birmingham on 18th January 1967.

Davies, R.W., "Research, Development and Innovation in the Soviet Economy, 1968-1970", in D.O. Edge and J.N. Wolfe (eds.), Meaning and Control. Social Aspects of Science and Technology (London: Tavistock, 1973), 241-261.

Davies, R.W. and R. Amann, "Science Policy in the USSR", Scientific American, No. 6 (1969), 19-29.

Davies, R.W. and M.J. Berry, "The Research and Development Effort in the Soviet Union: A Reconsideration of the Manpower Data", in Science Policy in the USSR (Paris: OECD, 1969), 501-557.

Die DDR - ein moderner Industriestaat (Dresden: Verlag Zeit im Bild, 1966).

Die DDR - Entwicklung, Probleme, Perspektiven (Frankfurt am Main: Verlag Marxistische Blätter, 1972).

DDR Handbuch (Köln: Verlag Wissenschaft und Politik, 1979).

Die Deutsche Akademie der Wissenschaften auf dem Weg zur Forschungsakademie der sozialistischen Gesellschaft, Schriftenreihe des Staatsrates, Heft 12 (Berlin: Staatsverlag der Deutschen Demokratischen Republik, 1970).

Deutsches Institut für Wirtschaftsforschung, DDR Wirtschaft. Eine Bestandsaufnahme (Frankfurt am Main: Fischer Verlag, 1974).

Deutsches Institut für Wirtschaftsforschung, Handbuch DDR Wirtschaft (Reinbek bei Hamburg: Rowohlt Verlag, 1977).

Deutsches Institut für Wirtschaftsforschung, "Das Sozialprodukt der DDR und der Bundesrepublik Deutschland im Vergleich", Wochenbericht, Nr. 23-24 (1977), 195-200.

Dewey, C., "Ökonomischer Nutzen und Planvorbereitung 1963", Deutsche Finanzwirtschaft, 16, Nr. 24 (1962), 9-13.

Dinnies, G. and R. Köhler, "Wirksamere Preisbestimmungen für neu- und weiterentwickelte sowie für veraltete Erzeugnisse der mvI", Die Wirtschaft, Nr. 1 (1971), 19-20.

Directives for the Five-Year Plan for the GDR's National Economic Development 1976-1980, issued by the 9th Congress of the Socialist Unity Party of Germany (Dresden: Verlag Zeit im Bild, 1976).

Direktive für den zweiten Fünfjahrplan zur Entwicklung der Volkswirtschaft der Deutschen Demokratischen Republik 1956 bis 1960 (Berlin: Dietz Verlag, 1956).

"Direktorenkonferenz der Forschungsgemeinschaft", Spektrum, Nr. 6 (1964), 241-249.

Dittrich, R. and J. Steiner, "Erfahrungen bei der Organisation des sozialistischen Wettbewerbs in produktionsvorbereitenden Bereichen", *Sozialistische Arbeitswissenschaft*, Nr. 5 (1973), 345-352.

Dokumente der Sozialistischen Einheitspartei Deutschlands, Band II (Berlin: Dietz Verlag, 1950).

Dokumente der Sozialistischen Einheitspartei Deutschlands, Band III (Berlin: Dietz Verlag, 1952).

Döring, W., G. Feldmann and W. Riede, *Der Plan Neue Technik. Planteil TOM und der Meister* (Berlin: Verlag Die Wirtschaft, 1961).

Dory, J.P. and R.J. Lord, "Does TF Really Work", *Harvard Business Review* (November-December 1970), 16-28, 168.

Dost, A., "Einige Probleme der Preisbildung zur Erhöhung der ökonomischen Wirksamkeit von Preis und Gewinn", *Die Wirtschaft*, Nr. 44 (1962), 11.

Dreyer, E., "Zur Anwendung des Prinzips der materiellen Interessiertheit in der Forschung und Entwicklung", *Die Wirtschaft*, Nr. 21 (1967), 9.

Durch sozialistische Gemeinschaftsarbeit zum wissenschaftlich-technischen Höchststand im Maschinenbau und in der Metallurgie. Materialien der 9. Tagung des ZK der SED (Berlin: Dietz Verlag, 1960).

"Die Durchführung des Volkswirtschaftsplanes im Jahre 1970", *Neues Deutschland* (11. Juni 1970), 3-6.

Ebel, H.,"Die Akademiereform verlangt politisch-ideologische Klarheit", *Spektrum*, Nr. 11 (1968), 373-375.

Echterhoff-Severitt, H., "Wissenschaftsaufwendungen in der Bundesrepublik Deutschland. Folge 2: FuE-Personal in Unternehmen und Verbänden", *Wirtschaft und Wissenschaft*, Nr. 2 (1969), 21-24.

Echterhoff-Severitt, H., "Wissenschaftsaufwendungen in der Bundesrepublik Deutschland. Folge 5: FuE-Personal in den Unternehmen und Verbänden im Jahre 1969", *Wirtschaft und Wissenschaft*, Nr. 6 (1971), XVII-XX.

Echterhoff-Severitt, H., *Forschung und Entwicklung in der Wirtschaft 1971* (Essen: Stifterverband für die deutsche Wissenschaft, Arbeitsschrift C, 1974).

Eggert, K., "Enge Bindung von Wissenschaft und Praxis, Diskussionsthema der Rektorenkonferenz der DDR", *Das Hochschulwesen*, Nr. 7-8 (1959), 363-364.

Eggert, K., "Hemmnisse in der Zusammenarbeit Universität - Industrie überwinden", *Das Hochschulwesen*, Nr. 7 (1964), 463-469.

"Eine alte Kiste", *Die Wirtschaft*, Nr. 48 (1965), 17.

Emmerich, H., "Der Plan 'Neue Technik' erfordert eine höhere Qualität der Leitungstätigkeit", *Einheit*, Nr. 5 (1961), 693-704.

Emmerich, H., "Zur Methodik des Planes 'Neue Technik' 1962", *Sozialistische Planwirtschaft*, Nr. 7 (1961), 19-23.

Emmerich, H., "Für die zentrale Planung des technischen Fortschritts", *Sozialistische Planwirtschaft*, *3*, Nr. 5 (1962), 3-6.

Emmrich, W., "Was die Preisbildung an den Tag bringt", Die Wirtschaft, Nr. 12 (1962), 5.
"Empfehlung zur weiteren Entwicklung und Verbesserung der Arbeit der Deutschen Akademie der Wissenschaften zu Berlin", Mitteilungsblatt der Deutschen Akademie der Wissenschaften zu Berlin, Nr. 4-5 (1955), 1-27.
"Entschließung des 5. Plenums des Zentralkomitees der SED", Neues Deutschland, Ausgabe B (26.5.1959), 3.
"Die Entwicklung der Volkswirtschaft im Jahre 1970, dem letzten Jahr des Perspektivplans 1966 bis 1970", Neues Deutschland (22. Januar 1971), 3-4.
"Entwicklung des wissenschaftlich-technischen Potentials der DDR", Informatik, Nr. 4 (1972), 9.
"Entwicklung des wissenschaftlich-technischen Potentials der DDR", Die Wirtschaft, Nr. 7 (1972), 20.
"Erkenntnisse aus der Erfüllung des Planes Wissenschaft und Technik", Die Wirtschaft, Nr. 21 (1974), 2.
Facts and Figures about the Progress of the German Democratic Republic from 1971 to 1975 (Compiled by departments of the Central Committee of the SED and the Central Statistical Board of the GDR, 1976).
Falk, W., G. Richter and W. Schmidt, Wirtschaft - Wissenschaft - Welthöchststand (Berlin: Verlag Die Wirtschaft, 1969).
"Falsche Festpreise hemmen die Produktivität", Die Wirtschaft, Nr. 50 (1961), 5.
Feldmann, G. and H. Brottke, "Praxisnahe Berechnung des ökonomischen Nutzeffektes der Forschung und Entwicklung", Die Wirtschaft, Nr. 45 (1967), 10.
Feller, R. and S. Strauß, "Fonds Wissenschaft und Technik als Leitungsinstrument", Die Wirtschaft, Nr. 5 (1970), 5.
Fiedler, P. and G. Riege, "Das Anstrengendste steht noch bevor", Forum, Nr. 2 (1970), 8-9, 18.
Der Fischer Weltalmanach 1981 (Frankfurt am Main: Fischer Verlag, 1980).
Forschung und Technik, Schriftenreihe Der Fünfjahrplan, Heft 2 (Berlin: Verlag Die Wirtschaft, 1952).
"Forschungsinstitute nutzen ihre Aggregate gemeinsam", Neues Deutschland (11.2.1976), 2.
"Forschungskapazität vertraglich gebunden", Die Wirtschaft, Nr. 16 (1965), 4.
"Forschungsplanung mit Risiko?", Die Wirtschaft, Nr. 18 (1974), 7.
"Forschungsrat tagte vor dem 5. Plenum des ZK", Die Wirtschaft, Nr. 1 (1964), 4.
Foth, E., "Produktion neuer Erzeugnisse genügend stimuliert?", Die Wirtschaft, Nr. 39 (1967), 14.
Freeman, C., "Technology Assessment and its Social Context", Studium Generale, 24 (1971), 1038-1050.
Freeman, C., The Economics of Industrial Innovation (Harmondsworth: Penguin Books, 1974).
Freeman, C., "Economics of Research and Development", in I. Spiegel-Rösing and D. de Solla-Price (eds.), Science, Technology and Society. A Cross-Disciplinary Perspective (London: Sage, 1977), 223-275.

Freeman, C., Innovation and Size of Firm, Occasional Papers No 1 (Science Policy Research Centre, Griffith University, March 1978).

Freeman, C., "The Determinants of Innovation", Futures (June 1979), 206-215.

Freeman, C. and A. Young, The Research and Development Effort in Western Europe, North America and the Soviet Union (Paris: OECD, 1965).

Freeman, C., C. Cooper and K. Pavitt, "Policies for Technical Change", in C. Freeman and M. Jahoda (eds.), World Futures (London: Martin Robertson, 1978), 207-229.

Friedrich, G. and W. Schulz, "Zu einigen Aufgaben der VVB bei der Leitung des technischen Fortschritts in der Industrie", Einheit, Nr. 6 (1962), 33-44.

Friedrich, W., "Interview der 'Wirtschaft' mit dem Ersten Vertreter des Präsidenten der Deutschen Akademie der Wissenschaften zu Berlin", Mitteilungsblatt für die Mitarbeiter der Deutschen Akademie der Wissenschaften zu Berlin, Nr. 1-4 (1956), 31-32.

Fritzsche, H., "Mehr materiellen Anreiz bei betrieblichen Forschungs- und Entwicklungsarbeiten", Deutsche Finanzwirtschaft, Nr. 2 (1962), F9-F12.

Fritzsche, H., "Aufgaben der Technologie nach dem 9. Plenum", Die Wirtschaft, Nr. 47 (1968), 3.

Fülle, H., Der Absatz der Erzeugnisse unserer sozialistischen Industrie (Berlin: Volk und Wissen Volkseigener Verlag, 1961).

Garbe, E., "Reserven in knappen Konstruktionskapazitäten", Die Wirtschaft, Nr. 38 (1966), 12.

Garbe, E., "Brauchen wir 'Änderungs-Reserven'?", Die Wirtschaft, Nr. 49 (1972), 7.

"GDR Engineering Week in Great Britain" (Advertisement), Financial Times (10 November 1975), 8-11.

GDR Facts and Figures (Dresden: Verlag Zeit im Bild, various years).

GDR - 100 Questions 100 Answers (Berlin: Panorama DDR - Auslandspresseagentur, 2nd edition 1975).

Gericke, R. and W. Heerdegen, "Probleme der Anwendung des Auftragleitersystems in Forschung und Entwicklung", Die Technik, Nr. 10 (1970), 641-644.

German Democratic Republic. Industry (Dresden: Verlag Zeit im Bild, 1972).

"Gesetz über den Siebenjahrplan zur Entwicklung der Volkswirtschaft der Deutschen Demokratischen Republik in den Jahren 1959 bis 1965", in W. Ulbricht, Der Siebenjahrplan des Friedens, des Wohlstandes und des Glücks (Berlin: Dietz Verlag, 1959), 157-312.

Gesetzblätter der DDR (Arranged according to year, part and number):

"Bekanntmachung des Beschlusses des Ministerrats über Maßnahmen zur Förderung des wissenschaftlich-technischen Fortschritts in der DDR", GBl, T I, Nr. 63 (1955), 521-531.

"Anordnung über die Finanzierung und Verrechnung der Forschungs- und Entwicklungsarbeiten in den Betrieben der volkseigenen Wirtschaft", GBl, T I, Nr. 82 (1957), 685.

"Anordnung Nr. 2 über die Finanzierung und Verrechnung der Forschungs- und Entwicklungsarbeiten in den Betrieben der volkseigenen Wirtschaft vom 28. April 1958", GBl, T I, Nr. 32 (1959), 526.

"Anordnung über die Gewährung von Gewinnzuschlägen vom 28. April 1959", GBl, T I, Nr. 32 (1959), 526-527.

"Anordnung Nr. 2 über die Gewährung von Gewinnzuschlägen vom 21. März 1960", GBl, T I, Nr. 22 (1960), 223-224.

"Anordnung Nr. 3 über die Finanzierung und Verrechnung der Forschungs- und Entwicklungsarbeiten in den Betrieben der volkseigenen Wirtschaft vom 21. März 1960", GBl, T I, Nr. 22 (1960), 224-225.

"Anordnung über die Abführung von Gewinnabschlägen zur weiteren Durchsetzung des wissenschaftlich-technischen Fortschritts vom 18. Dezember 1961", GBl, T III, Nr. 34 (1961), 399-401.

"Beschluß über die Ausarbeitung und Anwendung von Betriebsprämienordnungen in den volkseigenen und ihnen gleichgestellten Betrieben", GBl, T II, Nr. 14 (1962), 119-122.

"Richtlinie für das neue ökonomische System der Planung und Leitung der Volkswirtschaft", GBl, T II, Nr. 64 (1963), 453-498.

"Anordnung über die vorläufige Regelung zur Bildung und Verwendung des Fonds Technik in den dem Volkswirtschaftsrat unterstehenden Vereinigungen Volkseigener Betriebe für das Jahr 1964", GBl, T II, Nr. 89 (1963), 703-706.

"Beschluß über die Bildung und Verwendung des einheitlichen Prämienfonds in den volkseigenen und ihnen gleichgestellten Betrieben der Industrie und des Bauwesens und in den VVB im Jahre 1964 vom 30. Januar 1964", GBl, T II, Nr. 10 (1964), 80-86.

"Verordnung über die Preisbildung nach der Güteklassifizierung des Deutschen Amtes für Meßwesen und Warenprüfung. - Preisbildungsverordnung, Güteklassifizierung - vom 29. Januar 1964", GBl, T II, Nr. 14 (1964), 117-118.

"Anordnung über die Gewährung von Gewinnzuschlägen und über die Beauflagung von Gewinnabschlägen vom 11. Februar 1964", GBl, T III, Nr. 15 (1964), 158-160.

"Beschluß über die Grundsätze für die Bildung und Verwendung des einheitlichen Prämienfonds in der volkseigenen Wirtschaft im Jahre 1965", GBl, T II, Nr. 42 (1965), 297-299.

"Anordnung zur Bildung und Verwendung des Fonds Technik", GBl, T III, Nr. 26 (1965), 125-128.

"Beschluß zur Richtlinie für die Bildung und Verwendung des Prämienfonds in den volkseigenen und ihnen gleichgestellten Betrieben und den VVB der Industrie und des Bauwesens im Jahre 1967 sowie zur Übergangsregelung für das Jahr 1966 vom 7. April 1966", GBl, T II, Nr. 40 (1966), 249-256.

"Anordnung über die Preisbildung für neu- und weiterentwickelte sowie für veraltete Erzeugnisse der metallverarbeitenden Betriebe", GBl, T II, Nr. 64 (1967), 423-428.

"Beschluß über die Grundsatzregelung für komplexe Maßnahmen zur weiteren Gestaltung des ökonomischen Systems des Sozialismus in der Planung und Wirtschaftsführung für die Jahre 1969 und 1970 vom 26. Juni 1968", GBl, T II, Nr. 66 (1968), 433-452.

"Verordnung über die Bildung und Verwendung des Prämienfonds in den volkseigenen und ihnen gleichgestellten Betrieben, volkseigenen Kombinaten, den VVB (Zentrale) und Einrichtungen für die Jahre 1969 und 1970 vom 26. Juni 1968", GBl, T II, Nr. 67 (1968), 490-493.

"Richtlinie zur Einführung des fondsbezogenen Industriepreises und der staatlichen normativen Regelung für die planmäßige Senkung von Industriepreisen in den Jahren 1969/1970 vom 26. Juni 1968", GBl, T II, Nr. 67 (1968), 497-504.

"Anordnung über die auftragsgebundene Finanzierung wissenschaftlich-technischer Aufgaben und die Bildung und Verwendung des Fonds Wissenschaft und Technik vom 30. September 1968", GBl, T II, Nr. 110 (1968), 859-865.

"Richtlinien über die Preisbildung für wissenschaftlich-technische Leistungen vom 30. September 1968", GBl, T II, Nr. 110 (1968), 865-867.

"Verordnung über die Bildung und Rechtsstellung von volkseigenen Kombinaten", GBl, T II, Nr. 121 (1968), 963-965.

"Verordnung über das Verfahren der Gründung und Zusammenlegung von volkseigenen Betrieben", GBl, T II, Nr. 121 (1968), 965-968.

"Anordnung Nr. Pr. 12 über die Preisreformen bei Industriepreisen vom 14. November 1968", GBl, T II, Nr. 122 (1968), 971-973.

"Anordnung über die Preisbildung für neue und weiterentwickelte sowie veraltete Erzeugnisse der chemischen Industrie", GBl, T II, Nr. 122 (1968), 977-981.

"Anordnung über die Preisbildung für Gußerzeugnisse, die nach neu- und weiterentwickelten sowie veralteten Fertigungsverfahren oder Gußwerkstoffen hergestellt werden", GBl, T II, Nr. 9 (1969), 83-87.

"Anordnung über die Planung, Leitung und Finanzierung von wissenschaftlich-technischen Aufgaben der Universitäten, Hoch- und Fachschulen", GBl, T II, Nr. 16 (1969), 117-123.

"Anordnung über die Bildung und Verwendung des Prämienfonds sowie des Kultur- und Sozialfonds in naturwissenschaftlich-technischen Forschungseinrichtungen der Deutschen Demokratischen Republik vom 14. Februar 1969", GBl, T II, Nr. 20 (1969), 142-143.

"Beschluß über die Grundsätze für die Gestaltung des Auftragleitersystems für wirtschaftlich entscheidende Aufgaben", GBl, T II, Nr. 27 (1970), 197-199.

"Anordnung Nr. 3 über die Preisbildung für neu- und weiterentwickelte sowie für veraltete Erzeugnisse der metallverarbeitenden Betriebe vom 28. Mai 1970", GBl, T II, Nr. 55 (1970), 417-418.

"Beschluß über die Durchführung des ökonomischen Systems des

Sozialismus im Jahre 1971 vom 1. Dezember 1970", GBl, T II, Nr. 100 (1970), 731-746.

"Verordnung über die Planung, Bildung und Verwendung des Prämienfonds und des Kultur- und Sozialfonds für das Jahr 1971", GBl, T II, Nr. 16 (11. Februar 1971), 105-111.

"Beschluß über Maßnahmen auf dem Gebiet der Leitung, Planung und Entwicklung der Industriepreise vom 17. November 1971", GBl, T II, Nr. 77 (1971), 669-673.

"Verordnung über die Planung, Bildung und Verwendung des Prämienfonds und des Kultur- und Sozialfonds für volkseigene Betriebe im Jahre 1972", GBl, T II, Nr. 5 (1972), 49-53.

"Anordnung über die Planung, Bildung und Verwendung des Leistungsfonds der volkseigenen Betriebe", GBl, T II, Nr. 42 (1972), 467-469.

"Verordnung über die Leitung, Planung und Finanzierung der Forschung an der Akademie der Wissenschaften und an Universitäten und Hochschulen vom 23. August 1972", GBl, T II, Nr. 53 (1972), 589-594.

"Anordnung über die zentrale staatliche Kalkulationsrichtlinie zur Bildung von Industriepreisen", GBl, T II, Nr. 67 (1972), 741 ff.

"Anordnung Nr. 2 über die Planung, Bildung und Verwendung des Leistungsfonds der volkseigenen Betriebe", GBl, T I, Nr. 7 (1974), 66.

"Anordnung über die zentrale staatliche Kalkulationsrichtlinie zur Bildung von Industriepreisen", GBl, T I, Nr. 24 (1976), 321-345.

Gibbons, M. and R. Johnston, "The Roles of Science in Technological Innovation", Research Policy, 3 (1974), 220-242.

Gierisch, C., "Die neuen Bestimmungen über Planung, Vorbereitung, Durchführung und Finanzierung der Investitionen und die Aufgaben der Finanzorgane", Deutsche Finanzwirtschaft, Nr. 3 (1963), 4-8.

Gießmann, E.-J., "Bildungswesen bestimmt maßgeblich Entwicklungstempo", in W. Ulbricht, Das neue ökonomische System der Planung und Leitung der Volkswirtschaft in der Praxis (Berlin: Dietz Verlag, 1963), 241-260.

Gläser, K., "Das Institut für Werkzeugmaschinen Karl-Marx-Stadt", Die Technik, Nr. 10 (1961), 682-687.

Gluschkow, W. and G. Dobrow, "Die wissenschaftliche Prognose", Die Wirtschaft, Nr. 3 (1969), 30.

Grabis, J., W. Liebig and J. Weiß, "Forschungsorganisation im Kombinat", Die Wirtschaft, Nr. 5 (1971), 13-14.

Graichen, D. and L. Rouscik, Zur sozialistischen Wirtschaftsorganisation. Aufgaben - Probleme - Lösungen (Berlin: Verlag Die Wirtschaft, 1971).

Graichen, D. and B. Siegert, Sozialistische Wirtschafts-, Wissenschafts- und Leitungsorganisation (Berlin: Verlag Die Wirtschaft, 1974).

Gramsch, H.-U., "Durchsetzung sozialistischer Großforschung", Die Wirtschaft, Nr. 14 (1969), 3.

Granick, D., "Variations in Management of the Industrial Enterprise in Socialist Eastern Europe", in Reorientation and Commercial Relations of the Economies of Eastern Europe, A Compendium of Papers submitted to the Joint Economic Committee Congress of the United States (Washington, D.C.: U.S. Government Printing Office, 1974), 229-247.

Granick, D., Enterprise Guidance in Eastern Europe. A Comparison of Four Socialist Economies (Princeton, N.J.: Princeton University Press, 1975).

Gregory, P. and G. Leptin, "Similar Societies under Differing Economic Systems: The Case of the Two Germanies", Soviet Studies, No. 4 (1977), 519-542.

Gripinski, L., "Kooperationsbeziehungen zwischen Hochschule und Industrie", Das Hochschulwesen, Nr. 4 (1969), 244-250.

Groschupf, H., "Wissenschaft und Hochschulbildung im Dienste des gesellschaftlichen Fortschritts", Das Hochschulwesen, Nr. 10 (1974), 292-295.

Grosse, H., "Für eine höhere Qualität der Pläne 'Neue Technik'", Die Wirtschaft, Nr. 40 (1960), 7.

Grote, C., "Keine Pionier- und Spitzenleistungen ohne gesellschaftlichen Nutzen", Spektrum, Nr. 1 (1971), 21-23.

Gruhn, W., "Zur Industrieforschung in der DDR. Teil I: Grundsätzliche Probleme", Deutsche Studien, Nr. 44 (1973), 368-377.

Gruhn, W., "Zur Industrieforschung in der DDR. Teil II: Die Struktur der Industrieforschung", Deutsche Studien, Nr. 45 (1974), 64-83.

"Grundlagenforschung auch in der Industrie", Die Wirtschaft, Nr. 12 (1974), 7.

"Grundsatzregelung für die Gestaltung des ökonomischen Systems des Sozialismus in der Deutschen Demokratischen Republik im Zeitraum 1971-1975", Die Wirtschaft, Beilage 14 zur Ausgabe Nr. 18 (29. April 1970).

Hager, K., "Hochschulen werden zu Kombinaten der Wissenschaft", Die Wirtschaft, Nr. 15 (1969), 4.

Hager, K., "Neue Etappe unserer Wissenschaftspolitik", Neues Deutschland (5. April 1969), 4.

Hager, K., Sozialismus und wissenschaftlich-technische Revolution (Berlin: Dietz Verlag, 2. Auflage 1973).

Hager, K., Wissenschaft und Technologie im Sozialismus (Berlin: Dietz Verlag, 1974).

Hamberg, D., "Invention in the Industrial Research Laboratory", Journal of Political Economy (April 1963), 95-115.

Hanhardt, A.M., Jr., "Die Ordentlichen Mitglieder der Deutschen Akademie der Wissenschaften zu Berlin 1945-1961", in P.C. Ludz (ed.), Studien und Materialien zur Soziologie der DDR (Köln und Opladen: Westdeutscher Verlag, 1964), 241-262.

Hanicke, R., "Wissenschaftlich-technischer Fortschritt, Weltniveau und Höchststand", Einheit, Nr. 7 (1964), 128.

Harig, G. and H. Neels (eds.), Die Entwicklung der Wissenschaft zur unmittelbaren Produktivkraft, Materialien der wissenschaftlichen Tagung des Prorektorats für Forschungsangelegenheiten der Karl-Marx-Universität zum 10. Jahrestag ihrer Namensgebung am 3. und 4. Mai 1963 (Leipzig: Karl-Marx-Universität, 1963).

Hartke, W., "Leibniz-Tag, 30. Juni 1966", in Jahrbuch der Deutschen Akademie der Wissenschaften zu Berlin (1966), 77-98.

Hartke, W., "Hauptversammlung der Akademie", in Jahrbuch der Deutschen Akademie der Wissenschaften zu Berlin (1967), 110-131.

Hartkopf, W. and G. Dunken, Von der Brandenburgischen Sozietät der Wissenschaften zur Deutschen Akademie der Wissenschaften zu Berlin (Berlin: Akademie Verlag, 1967).

Hartmann, K., "Zur sozialistischen Konzentration und Kooperation", Die Wirtschaft, Nr. 14 (1968), 16-17.

Hasler, E., "Praktische Probleme der Prognosearbeit", Die Wirtschaft, Nr. 33 (1968), 8.

Haustein, H.-D. and K. Neumann, Die ökonomische Analyse des technischen Niveaus der Industrieproduktion. Teil I (Berlin: Verlag Die Wirtschaft, 1965).

Haustein, H.-D. and K. Neumann, Die ökonomische Analyse des technischen Niveaus der Industrieproduktion. Teil II (Berlin: Verlag Die Wirtschaft, 1965).

Heinelt, M., G. Klose and J. Zobel, "Plan und Leistungsbewertung müssen den technischen Fortschritt besser stimulieren", Die Wirtschaft, Nr. 18 (1974), 8.

Heinze, R., "Wissenschaftlich-technische Grundkonzeptionen schaffen", Die Wirtschaft, Nr. 39 (1964), 15-16.

Hemmerling, J., "Mit den Neuerungen zum wissenschaftlich-technischen Höchststand", Die Technik, Nr. 11 (1963), 704-705.

Hennig, G. and G. Reinecke, "Nutzen und Selbstkostensenkung müssen eins sein", Deutsche Finanzwirtschaft, Nr. 15 (1963), F9-F13.

Henschel, W., "Die Neue Technik darf nicht zu teuer sein", Die Wirtschaft, Nr. 22 (1961), 6.

Herber, R. and H. Jung, Wissenschaftliche Leitung und Entwicklung der Kader (Berlin: Staatsverlag der Deutschen Demokratischen Republik, 2. Auflage 1964).

Herpich, H., "Grundmittelanalyse an einer Universität", Sozialistische Finanzwirtschaft, Nr. 8 (1972), 48-50.

Herzog, H. and J. Storch, Der Meister und das Erfindungs- und Vorschlagswesen (Berlin: Verlag Die Wirtschaft, 1963).

Herzog, H. and J. Storch, Der Meister und die Neuererbewegung (Berlin: Verlag Die Wirtschaft, 1965).

Heuer, K., "Die wirtschaftliche Rechnungsführung in den naturwissenschaftlich-technischen Instituten der sozialistischen Industrie und die Gewährleistung einer neuen Qualität der Planung, Leitung, Organisation und Wirtschaftsführung im wissenschaftlich-technischen Zentrum der VVB", Wissenschaftliche Zeitschrift der Humboldt-Universität zu Berlin, Mathematisch-Naturwissenschaftliche Reihe, Nr. 4 (1964), 553-561.

Heuer, R. and H. Marx, "Zur Messung der Arbeitsproduktivität", Die Arbeit, Nr. 4 (1973), 32-34.

Heyde, S., "Die Messung der Arbeitsproduktivität und die Praxis", Wirtschaftswissenschaft, Nr. 11 (1962), 1679-1687.

Heyden, G., A. Kosing, O. Reinhold and H. Ullrich (Redaktion), Sozialismus-Wissenschaft-Produktion. Über die Rolle der Wissenschaft beim umfassenden Aufbau des Sozialismus in der DDR (Berlin: Dietz Verlag, 1963).

Hicke, H.-J., "Technika intensivieren Forschungsarbeit", Der Neuerer, Nr. 3 (1976), 98.

"Die historische Leistung W.I. Lenins auf dem Gebiet der Wirtschaftstheorie und Wirtschaftspolitik", Die Wirtschaft, Beilage zur Ausgabe Nr. 45 (1969), 4-7.

Hoche, F. and H. Schäfer, "Beziehungen zwischen Hochschulen und Praxispartnern in der naturwissenschaftlich-technischen Forschung", Wirtschaftsrecht, 1, Nr. 1 (1973), 18-21.

"Höchstleistungen und sozialistische Großforschung", Die Wirtschaft, Nr. 8 (1969), 2.

Hofmann, U., "Systematischer Ausbau der technisch-technologischen Basis", Spektrum, Nr. 10 (1975), 8-13.

Hofmann, U., "Wissenschaftspolitik und Forschungstechnologie", Spektrum, Nr. 6 (1976), 10-12.

Hofmann, U. and S. Heinrich, "Zur Zusammenarbeit zwischen der Akademie der Wissenschaften der DDR und dem Ministerium für Hoch- und Fachschulwesen", Das Hochschulwesen, Nr. 11 (1976), 307-311.

"Höhere Qualitätsziele im Plan 1975", Die Wirtschaft, Nr. 34 (1974), 2.

Honecker, E., Bericht des Zentralkomitees an den VIII. Parteitag der Sozialistischen Einheitspartei Deutschlands, Berlin, 15. bis 19. Juni 1971 (Berlin: Dietz Verlag, 9. Auflage 1975; 1. Auflage 1971).

Honecker, E., "Bericht des Zentralkomitees der Sozialistischen Einheitspartei Deutschlands an den IX. Parteitag der SED", Neues Deutschland (19. Mai 1976), 3-15.

Hornich, S., "Die Pläne Wissenschaft und Technik zum wichtigsten Bestandteil der Volkswirtschaftspläne machen", Die Wirtschaft, Nr. 29 (1974), 3-4.

Höstermann, H., "Die Preisreformen in der Industrie", Deutsche Finanzwirtschaft, Nr. 11 (1967), 10-11.

Immig, G., "Gewinnabschläge helfen die neue Technik durchsetzen und werden das Produktionsniveau heben", Die Wirtschaft, Nr. 2 (1962), 3.

"In größerem Umfang Pionier- und Spitzenleistungen erzielen", Die Wirtschaft, Nr. 37 (1969), 2.

"Innovation. Ideen zum Erfolg", Wirtschaftswoche, Nr. 17 (21. April 1978).

Institut für Gesellschaft und Wissenschaft Erlangen, abg, Analysen und Berichte aus Gesellschaft und Wissenschaft.

Institut für Gesellschaft und Wissenschaft Erlangen, igw Informationen. Zur Wissenschaftsentwicklung und -politik in der DDR.

Institut für Gesellschaft und Wissenschaft Erlangen (ed.), Wissenschaft in der DDR (Köln: Verlag Wissenschaft und Politik, 1973).

Institut für Gesellschaft und Wissenschaft Erlangen, "Zum 275. Gründungstag der Akademie der Wissenschaften", abg, Nr. 5 (1975).

Institut für Gesellschaftswissenschaften beim Zentralkomitee der SED (ed.), Wissenschaft und Produktion im Sozialismus (Berlin: Dietz Verlag, 1976).

Introducing the GDR (Dresden: Verlag Zeit im Bild, fourth, revised edition 1973).

"Ist unser wissenschaftlicher Vorlauf ausreichend?", Die Wirtschaft, Nr. 15 (1974), 4.

Jablonski, "Eine Verordnung ist noch kein Hebel", Die Wirtschaft, Nr. 29 (1970), 5.

"10 Jahre Deutsches Brennstoffinstitut", Neues Deutschland (5. Oktober 1966), 2.

Jewkes, J., D. Sawers and R. Stillerman, The Sources of Invention (London: Macmillan, first edition 1959, second edition 1969).

Jonuscheit, K.-H., "Das ökonomische System der Planung und Leitung unserer Volkswirtschaft, das Prinzip des demokratischen Zentralismus und die materielle Interessiertheit", Einheit, Nr. 5 (1963), 59-68.

Junker, L., "Finanzierung der Ingenieurbüros nach Effektivität", Sozialistische Finanzwirtschaft, Nr. 6 (1970), 50-53.

Kalweit, W., "Prämien für moderne oder für veraltete Technik?", Neues Deutschland, Ausgabe A (15. November 1962), 5.

"Kampf gegen Planrückstände und Schaffung des wissenschaftlich-technischen Verlaufs gehören zusammen", Die Wirtschaft, Nr. 23 (1970), 3.

Kannengießer, L., Die Organisation der Beziehungen zwischen Wissenschaft und Produktion (Berlin: Staatsverlag der Deutschen Demokratischen Republik, 1967).

Kannengießer, L. and W. Meske, "Wirksamer Einsatz des Forschungspotentials", Spektrum, Nr. 10 (1972), 19-21.

Kehrer, G., "Wechselbeziehungen zwischen territorialer Produktions- und Wirtschaftsstruktur", in H. Richter (ed.), Beiträge zur territorialen Produktionsstruktur (Gotha/Leipzig: VEB Hermann Haack, Geographisch-Kartographische Anstalt, 1976), 49-60.

Kelm, G., Forschungsergebnis-Wirtschaftsvertrag-Produktion (Berlin: Staatsverlag der Deutschen Demokratischen Republik, 1964).

Kerda, H., "WTZ - seine Funktionen für den wissenschaftlich-technischen Fortschritt und die Rationalisierung", Die Wirtschaft, Nr. 23 (1966), 6-7.

Keren, M., "The New Economic System in the GDR: An Obituary", Soviet Studies, Nr. 4 (1972/73), 554-587.

Keren, M., "The Rise and Fall of the New Economic System", in L. Legters (ed.), The German Democratic Republic. A Developed Socialist Society (Boulder, Co.: Westview Press, 1978), 61-84.

Kiera, H.-G., Partei und Staat im Planungssystem der DDR. Die Planung in der Ära Ulbricht (Düsseldorf: Droste Verlag, 1975).

Killiches, H., "Höhere Effektivität in Forschung, Technik, Produktion", Das Hochschulwesen, Nr. 7 (1972), 210-212.

Klare, H., "Probleme und Maßnahmen für die weitere Entwicklung der Institute der Forschungsgemeinschaft", Spektrum, Nr. 4 (1962), 163-175.

Klare, H., "Forschung nach Plan und Konzeption", Neues Deutschland (9. Februar 1964), 3.

Klare, H., in Jahrbuch der Deutschen Akademie der Wissenschaften zu Berlin (1967), 139-149.

Klare, H., Interview in Wissenschaft und Produktion, Nr. 5 (1969), 22-25.

Klare, H., "Grundlagenforschung - Effektivität von übermorgen", Einheit, Nr. 10 (1972), 1283-1289.

Klare, H., "Wissenschaft im Auftrag und zum Nutzen unseres Volkes", Spektrum, Nr. 9 (1974), 5-11.

Klare, H., "Grundlagenforschung - Basis des wissenschaftlich-technischen Fortschritts", Einheit, Nr. 10 (1976), 1096-1101.

Klare, H. and H. Müller, "Grundlagenforschung für Wissenschaft und Produktion", Wissenschaft und Fortschritt, Nr. 5 (1976), 196-200.

Klare, H. and W. Hartkopf, Die Akademie der Wissenschaften der DDR. Zum 275. Jahrestag der Gründung der Akademie (Berlin: Akademie der Wissenschaften der DDR, 1975).

Klose, G., "Erfahrungen der Sowjetunion mit Koordinierungsplänen", Technische Gemeinschaft, Nr. 8 (1973), 21-23.

Klose, G. and M. Koss, "Durch sozialistische Forschungskooperation zu höherer Produktivität der wissenschaftlich-technischen Arbeit (Teil 1)", Die Technik, Nr. 5 (1972), 291-295.

Koch, H. and F. Paschke, "Erfahrungen aus der Arbeit mit Gegenplänen gehören in die Rahmenrichtlinie", Die Wirtschaft, Nr. 18 (1974), 9.

"Kombinatsbildung und Effektivität", Die Wirtschaft, Nr. 24 (1969), 11.

"Kommentare zum Entwurf der Grundsatzregelung für komplexe Maßnahmen zur weiteren Gestaltung des ökonomischen Systems des Sozialismus in der Planung und Wirtschaftsführung für die Jahre 1969 und 1970", Die Wirtschaft, Beilage zur Ausgabe Nr. 19 (1968).

"Kontinuierlich die Planaufgaben erfüllen", Die Wirtschaft, Nr. 24 (11. Juni 1970), 2.

"Konzentrationsprozeß", Die Wirtschaft, Nr. 32 (1968), 2.

Koschwitz, E., "Prämiierungsform in Industrieforschungsinstituten", Wirtschaftswissenschaft, Nr. 7 (1968), 1107-1121.

Koschwitz, E., "Diskussionsbeitrag", in A. Lange (ed.), Forschungsökonomie (Berlin: Verlag Die Wirtschaft, 1969), 240-244.

Kosta, H.G.J., H. Kramer and J. Slama, Der technologische Fortschritt in Österreich und in der Tschechoslowakei (Wien: Springer Verlag, 1971).

Krakat, K., "Der Weg zur dritten Generation. Die Entwicklung der EDV in der DDR bis zum Beginn der siebziger Jahre", FS Analysen, Nr. 7 (1976).

Kramer, W. and W. Linden, "Forschungskooperation MLU-VVB Agrochemie", Das Hochschulwesen, Nr. 11 (1969), 771-773.

Kratsch, O., "Zur sowjetischen Diskussion über 'Plan, Gewinn, Prämie'", Wirtschaftswissenschaft, Nr. 1 (1963), 109-126.

Krieger, H., "Probleme des Kaderaustausches Hochschule - Praxis", Das Hochschulwesen, Nr. 4-5 (1970), 295-299.

Kröber, G. and H. Laitko, Sozialismus und Wissenschaft (Berlin: VEB Deutscher Verlag der Wissenschaften, 1972).

Kröber, G. and H. Laitko, Wissenschaft als soziale Kraft (Berlin: Akademie Verlag, 1976).

Kuciak, G., "Ökonomische Gesetze und Betriebsgröße", Die Wirtschaft, Nr. 7 (1967), 11-12.

Kuciak, G., "Konzentration und optimale Betriebsgröße", Die Wirtschaft, Nr. 16 (1968), 13.

Kuczmera, M., "Die optimale Betriebsgröße - eine Notwendigkeit für unsere Volkswirtschaft", Die Wirtschaft, Nr. 7 (1967), 12.

Kuczmera, M., "Optimale Betriebsgröße und Arbeitsproduktivität in den Industriezweigen der DDR", Die Wirtschaft, Beilage zur Ausgabe Nr. 5 (1968), 12-13.

Kühnemund, R. and H.-J. Weihs, "Den wissenschaftlich-technischen Fortschritt auch mit Hilfe von Prämien beschleunigen", Arbeit und Arbeitsrecht, Nr. 16 (1976), 487-490.

"Kurze Überleitungsfristen garantiert Zeit-, Markt- und Produktivitätsgewinn", Die Wirtschaft, Nr. 2 (1966), 4-5.

Kusicka, H., "Probleme der Planung, Leitung und materiellen Stimulierung von Forschung und Entwicklung", in A. Lange (ed.), Forschungsökonomie (Berlin: Verlag Die Wirtschaft, 1969).

Kusicka, H. and W. Leupold, "Für eine höhere Effektivität der industriellen Forschung und Entwicklung", Einheit, Nr. 1 (1965), 20-28.

Kusicka, H. and W. Leupold, Industrieforschung und Ökonomie (Berlin: Dietz Verlag, 1966).

Laitko, H. and G. Wendel, "Wissenschaftspolitische Wandlungen", XV. Internationaler Kongreß für Geschichte der Wissenschaft, Edinburgh, August 1977.

Landrock, R., Die Deutsche Akademie der Wissenschaften zu Berlin 1945-1971, Band 1 (Erlangen: Deutsche Gesellschaft für zeitgeschichtliche Fragen e.V.).

Lange, A., "Den Plan 'Neue Technik' zum Hauptinstrument des wissenschaftlich-technischen Fortschritts machen!", Wissenschaftliche Zeitschrift der Hochschule für Ökonomie, Berlin, 7, Nr. 1 (1962), 1-6.

Lange, A., Ökonomie und neue Technik (Berlin: Verlag Die Wirtschaft, 1963).

Lange, A. (ed.), Forschungsökonomie (Berlin: Verlag Die Wirtschaft, 1969).

Lange, A., "Einige Probleme der sozialistischen Wissenschaftsorganisation", Wirtschaftswissenschaft, Nr. 9 (1969), 1281-1306.

Lange, A., "Den Zyklus 'Wissenschaft-Technik-Produktion' beherrschen (Teil I)", Die Technik, Nr. 1 (1973), 3-9.

Lange, A., "Ökonomische Probleme der Überführung wissenschaftlich-technischer Ergebnisse in die Produktion", Einheit, Nr. 1 (1974), 104-108.

Lange, A. and W. Marschall, "Ökonomische Probleme bei der Überführung wissenschaftlich-technischer Ergebnisse in die Produktion", Wirtschaftswissenschaft, Nr. 2 (1975), 195-208.

Lange, A. and D. Voigtberger, Überleitung von wissenschaftlich-technischen Ergebnissen (Berlin: Verlag Die Wirtschaft, 1975).

Lange, A., D. Ivanov, R. Zimmermann and P. Wieczorek, "Intensivierung der Produktion und komplexere Planung des wissenschaftlich-technischen Fortschritts", Die Technik, Nr. 7 (1976), 450-454.

Lange, H., "Die Konzentrationseffekte und ihre Realisierung, Grundlagen der höheren Effektivität der Kombinate", Wissenschaftliche Zeitschrift der Hochschule für Ökonomie, Berlin, Nr. 2 (1970), 111-118.

Langhoff, N., "Wissenschaftlicher Gerätebau an der Akademie", Spektrum, Nr. 11 (1976), 26-28.

Langrish, J., M. Gibbons, W.G. Evans and F.R. Jevons, Wealth from Knowledge. Studies of Innovation in Industry (London: Macmillan, 1972).

Lauter, E.A., "Wissenschaftsorganisation an der Akademie", Spektrum, Nr. 7 (1970), 6-7.

Leger, H. and I. Waltenberg, "Wissenschaftlich-technische Arbeit im Blickfeld der Parteikontrolle", Einheit, Nr. 11 (1975), 1304-1308.

Leisewitz, A. and R. Rilling, "Wissenschafts- und Forschungspolitik in BRD und DDR", in BRD - DDR Vergleich der Gesellschaftssysteme (Köln: Pahl-Rugenstein, 1971), 365-384.

"Leistungsaufgaben und -instrumente für die Vervollkommnung der Planung der wissenschaftlich-technischen Arbeit", Die Wirtschaft, Beilage zur Ausgabe Nr. 31 (1973), 5-13.

Lenin, W.I., "Was sind die 'Volksfeinde' und wie kämpfen sie gegen die Sozialdemokraten", in Werke, Band 1 (Berlin: Dietz Verlag, 1971), 119-338.

Lenin, W.I., "Ursprünglicher Entwurf des Artikels 'Die nächsten Aufgaben der Sowjetmacht'", in Werke, Band 27 (Berlin: Dietz Verlag, 1972), 192-208.

Lenin, W.I., "Entwurf eines Plans wissenschaftlich-technischer Arbeiten", in Werke, Band 27 (Berlin: Dietz Verlag, 1972), 312-313.

Lenin, W.I., "Rede auf dem I. Gesamtrussischen Kongress der

Volkswirtschaftsräte, 26. Mai 1918", in Werke, Band 27 (Berlin: Dietz Verlag, 1972), 404-412.

Lenin, W.I., "Rede auf dem II. Gesamtrussischen Verbandstag des Medizinischen und Sanitätspersonals, 1. März 1920", in Werke, Band 30 (Berlin: Dietz Verlag, 1972), 393-394.

Lenin, W.I., "Gesamtrussischer Verbandstag der Bergarbeiter", in Werke, Band 32 (Berlin: Dietz Verlag, 1961), 39-55.

Leptin, G. and M. Melzer, Economic Reforms in East German Industry (London: Oxford University Press, 1978).

Leupold, W., "Rationeller Einsatz der Mittel und Kräfte in Wissenschaft und Technik", Energietechnik, Nr. 4 (1973), 147-152.

Leuschner, B., "Aktuelle Probleme unserer Volkswirtschaft", Die Wirtschaft, Nr. 51 (1956), 5-6.

Lexikon der Wirtschaft. Industrie (Berlin: Verlag Die Wirtschaft, 1970).

Liebig, W., "Der Kampf der Betriebsparteiorganisation um den wissenschaftlich-technischen Fortschritt", Einheit, Nr. 10 (1962), 136-138.

Liebscher, F., "Die Technische Universität Dresden auf dem Wege zur sozialistischen Universität", Die Technik, Nr. 10 (1969), 649-653.

Lindblom, C.E., The Policy-Making Process (Englewood Cliffs, N.J.: Prentice-Hall, second edition 1980).

Linnemann, G. and S. Wykowski, "Erfahrungen und Zielstellungen im Zusammenwirken von Technischer Hochschule und Industrie auf dem Gebiet von Grundlagen- und angewandter Forschung", Das Hochschulwesen, Nr. 11 (1976), 304-307.

Lilie, H., "Aus den Erfahrungen der bisherigen Prognosearbeit", Die Wirtschaft, Nr. 21 (1970), 5-6.

Lorenz, H.-J. and R. Haker, "Die Planung, Bilanzierung und Realisierung volkswirtschaftlich strukturbestimmender Aufgaben", Die Wirtschaft, Nr. 30 (1968), 5.

Lücke, P.R., "Wissenschaft und Forschung im anderen Teil Deutschlands. Wissenschaftliche Akademien", Deutsche Studien, Nr. 32 (1970), 398-405.

Lysk, H., "Zielstrebige politisch-ideologische Arbeit beschleunigt wissenschaftlich-technischen Fortschritt", Einheit, Nr. 7 (1975), 714-718.

Maier, K. and G. Wagner, "Kampf den unvollendeten Investitionen und der Mittelzersplitterung", Deutsche Finanzwirtschaft, 16, Nr. 19 (1962), 3-5.

Mann, H., "Probleme der Preisbildung für neue Erzeugnisse (Neue Technik)", Deutsche Finanzwirtschaft, 16, Nr. 23 (1962), 3-6.

Mann, H., "Die planmäßige Preisbildung als Instrument zur Förderung des wissenschaftlich-technischen Fortschritts", Wirtschaftswissenschaft, Nr. 6 (1975), 826-842.

Mansfield, E., Research and Innovation in the Modern Cooperation (London: Macmillan, 1971).

Mansfield, E., "Technological Forecasting", in T.S. Khachaturov (ed.), Methods of Long-Term Planning and Forecasting (London: Macmillan, 1976), 334-349.

Marschall, W., "Ökonomische Probleme des wissenschaftlich-technischen Fortschritts", Einheit, Nr. 4 (1973), 496-499.

Marx, K., Das Kapital. Erster Band, in K. Marx and F. Engels, Werke, Band 23 (Berlin: Dietz Verlag, 1972).

Marx, K., Das Kapital. Dritter Band, in K. Marx and F. Engels, Werke, Band 25 (Berlin: Dietz Verlag, 1973).

Marx, K., Grundrisse der Kritik der politischen Ökonomie (Berlin: Dietz Verlag, 1974).

Melms, E., "Praktische Fragen der vorrangigen Planung, Bilanzierung und Realisierung volkswirtschaftlich strukturbestimmender Aufgaben", Die Wirtschaft, Nr. 37 (1968), 7.

Meske, W., "Entwicklungstendenzen in der Ausstattung der Forschung mit Grundmitteln, Teil I", Die Technik, Nr. 12 (1973), 748-751.

Meske, W., "Entwicklungstendenzen in der Ausstattung der Forschung mit Grundmitteln, Teil II", Die Technik, Nr. 1 (1974), 29-32.

Meske, W., "Zur Entwicklung des Forschungspotentials der DDR unter den Bedingungen der intensiv erweiterten Reproduktion im Sozialismus", in Sozialistische Wissenschaftspolitik und marxistisch-leninistische Wissenschaftstheorie, Kolloquien Reihe der Akademie der Wissenschaften der DDR, Institut für Wissenschaftstheorie und -organisation, Heft 10 (1975), 109-126.

Meyer, A., "Kompromisse bei der Kombinatsbildung zahlen sich nicht aus", Die Wirtschaft, Nr. 28 (1970), 5.

Meyer, J., "Methodik und Auswertung der Erfassung der Grundmittel und der Arbeitsfläche für Forschung und Entwicklung", Statistische Praxis, Nr. 7 (1967), 409-412.

Mittag, G., "Komplexe sozialistische Rationalisierung - eine Hauptrichtung unserer ökonomischen Politik bis 1970", in Sozialistische Rationalisierung und Standardisierung (Berlin: Dietz Verlag, 1966), 31-92.

Mittag, G., "Von der 3. Tagung des Zentralkomitees. Aus dem Bericht des Politbüros an das 3. Plenum", Neues Deutschland (24. November 1967), 2-8.

Mittag, G., "Meisterung der Ökonomie ist für uns Klassenkampf", Neues Deutschland (27. Oktober 1968), 5-6.

Mittag, G., "Unsere sozialistische Planwirtschaft ermöglicht hohe Effektivität", Die Wirtschaft, Beilage zur Ausgabe Nr. 18 (1968), 6-10.

Moc, N., "Lange Lieferzeiten bremsen das Tempo der Modernisierung", Die Wirtschaft, Nr. 49 (1966), 5.

Mulitze, H., Neuerungen - dokumentieren, anbieten, nachnutzen (Berlin: Staatsverlag der Deutschen Demokratischen Republik, 1966).

Müller, J., D. Schubert and B. Wilhelmi, "Überführung von Ergebnissen der physikalischen Forschung in die Produktion", Das Hochschulwesen, Nr. 10 (1974), 302-308.

Müller, K., "Phasen und Verkettungen der Überführung", Spektrum, Nr. 5 (1973), 16-18.

Müller, R. and Bulla, "Über die Zusammenarbeit des Institutes für Elasto- und Plastomechanik der TU Dresden mit der Industrie", Die Technik, Nr. 2 (1963), 68-72.

Müller, W., "Wie die Arbeitsproduktivität messen?", Die Wirtschaft, Nr. 13 (1971), 14.

Napierkowski, "Die Planung des Prämienfonds als absoluter Betrag", Die Wirtschaft, Nr. 13 (1971), 14.

The National Economy under Socialist Conditions. Objectives and Results in the GDR (Berlin: Panorama DDR - Auslandspresseagentur, 1974).

Naumann, F., "10 Jahre Institutsarbeit im Dienste der Gießereiindustrie", Die Technik, Nr. 7 (1963), 456-459.

"Neue Prämienordnung", Deutsche Finanzwirtschaft, Nr. 10 (1969), F1

Nick, H., Intensivierung und wissenschaftlich-technischer Fortschritt (Berlin: Dietz Verlag, 1974).

Nitze, G. and H.-J. Beckmann, "Der Plan Wissenschaft und Technik muß kontrollfähig sein", Die Wirtschaft, Nr. 20 (1974), 10.

Nolting, L.E., The Financing of Research, Development and Innovation in the USSR, by Type of Performer (Washington, D.C.: U.S. Department of Commerce, 1976).

"Nur ein modernes sozialistisches Planungssystem sichert Pionier- und Spitzenleistungen", Die Wirtschaft, Nr. 3 (1970), 3-4.

OECD, Trade by Commodities. Market Summaries. Imports January - December 1968, Series C (Paris: OECD).

OECD, A Study of Resources Devoted to R&D in OECD Member Countries in 1963/64, Vol. 2, Statistical Tables and Notes (Paris: OECD, 1968).

OECD, International Survey of the Resources Devoted to R&D in 1969 by Member Countries, Statistical Tables and Notes, Vol. I, Business Enterprise Sector (Paris: OECD, 1972).

OECD, International Survey of the Resources Devoted to R&D in 1969 by Member Countries, Vol. 3, Private Non-Profit Sector (Paris: OECD, 1972).

OECD, International Survey of the Resources Devoted to R&D in 1969 by OECD Member Countries, Statistical Tables and Notes, Vol. 5, Total Tables (Paris: OECD, 1973).

OECD, Survey of the Resources Devoted to R&D by OECD Member Countries, International Statistical Year 1971, Vol. 5, Total Tables (Paris: OECD, 1974).

OECD, Survey of the Resources Devoted to R&D by OECD Member Countries, International Statistical Year 1971, Vol. 1, Business Enterprise Sector (Paris: OECD, 1975).

OECD, Survey of the Resources Devoted to R&D by OECD Member Countries, International Statistical Year 1971, Vol. 3, Private Non-Profit Sector (Paris: OECD, 1975).

OECD, The Measurement of Scientific and Technical Activities. "Frascati Manual" (Paris: OECD, 1976).

OECD, Survey of the Resources Devoted to R&D by OECD Member Countries, International Statistical Year 1973, Vol. 5, Total Tables (Paris: OECD, 1976).

OECD, Survey of the Resources Devoted to R&D by OECD Member Countries, International Statistical Year 1973, Vol. 3, Private Non-Profit Sector (Paris: OECD, 1977).

OECD, Survey of the Resources Devoted to R&D by OECD Member Countries, International Statistical Year 1973, Vol. 4, Higher Education Sector (Paris: OECD, 1977).

Oldenbourg, F., "Zur Organisation der Forschung in der DDR. Bemerkung zur neuen 'Verordnung über die Leitung, Planung und Finanzierung an der Akademie der Wissenschaften und an Hochschulen' der DDR", Berichte des Bundesinstituts für Ostwissenschaftliche und Internationale Studien, Nr. 1 (1973).

Osers, F., Forschung und Entwicklung in sozialistischen Staaten Osteuropas (Berlin: Duncker und Humblot, 1974).

Pässler, E., "Die industrieverbundene Forschungsarbeit auf dem Gebiet der Fertigungstechnik im Maschinenbau. Aufgaben und Arbeitsweise des Zentralinstituts für Fertigungstechnik des Maschinenbaues (ZIF) in Karl-Marx-Stadt", Die Technik, Nr. 7 (1963), 451-455.

Pavitt, K., "Research, Innovation and Economic Growth", Nature, No. 4903 (1963), 1-12.

Pavitt, K., "Analytical Techniques in Government Science Policy", Futures (March 1972), 5-12.

Pavitt, K., "Science, Enterprise and Innovation", Minerva, XI, No. 2 (1973), 273-277.

Pavitt, K., "Government Policies Towards Innovation: A Review of Empirical Findings", Omega, No. 5 (1976), 539-558.

Pavitt, K., "The Choice of Targets and Instruments for Government Support of Scientific Research", in A. Whiting (ed.), The Economics of Industrial Subsidies (London: HMSO, 1976), 113-138.

Pavitt, K., "Technical Change. The Prospects for Manufacturing Industry", Futures (August 1978), 283-292.

Pavitt, K. (ed.), Technical Innovation and British Economic Performance (London: Macmillan, 1980).

Pavitt, K. and S. Wald, The Conditions for Success in Technological Innovation (Paris: OECD, 1971).

Pavitt, K. and W. Walker, "Government Policies Towards Industrial Innovation: A Review", Research Policy, 5 (1976), 11-97.

Pfützenreuter, W., "Die Verbesserung der Arbeit mit dem Plan Neue Technik", Deutsche Finanzwirtschaft, Ausgabe Finanzen und Buchführung, Nr. 12 (1962), F12-F13.

"Die Plandisziplin konsequent einhalten und festigen", Die Wirtschaft, Nr. 28 (9. Juli 1970), 2.

Pöschel, H., Leitung von Forschung und Technik und wissenschaftlicher Meinungsstreit (Berlin: Dietz Verlag, 1964).

Pöschel, H., "Forschung - Technik - Ideologie", Einheit, Nr 3 (1966), 302-311.

Pöschel, H., "Ökonomie und Ideologie in Forschung und Technik", Einheit, Nr. 8 (1969), 924-935.

Pöschel, H., "Ideologische Probleme der Leitung wissenschaftlich-technischer Arbeit", Einheit, Nr. 10 (1972), 1273-1282.

Pöschel, H. and S. Wikarski, "Durch ein System der einheitlichen und straffen Leitung zu einem hohen Nutzeffekt der wissenschaftlich-technischen Arbeit", Einheit, Nr. 2 (1962), 26-37.

Pöschel, H. and S. Wikarski, "Ideologische Probleme bei der Lösung der vom VI. Parteitag gestellten Aufgaben in Wissenschaft und Technik", Einheit, Nr. 6 (1963), 62-71.

Pohlisch, F., "Bilanz eines Jahres", Das Hochschulwesen, Nr. 2 (1968), 99-103.

Preispolitik und Preisbildung in der Deutschen Demokratischen Republik (Berlin: Verlag Die Wirtschaft, 1961).

Presse- und Informationsamt der Bundesregierung, Bulletin, Nr. 132 (10. Dezember 1970), 1263.

Prey, G., "Prognostik, Planung und Leitung auf dem Gebiet von Wissenschaft und Technik", Einheit, Nr. 10-11 (1967), 1353-1362.

Prey, G., "Auftragsgebunden", Deutsche Finanzwirtschaft, Nr. 15 (1968), 1, 15.

Prey, G., "Einige Führungsprobleme der sozialistischen Großforschung", Effekt, Nr. 3 (1969), 5-6.

Prey, G., "Zur Entwicklung der Großforschung im allgemeinen und der Aufbau der Großforschung im besonderen", Technische Gemeinschaft, Nr. 7 (1970), 8-9.

Prey, G., "Zur Entwicklung von Wissenschaft und Technik in der DDR bis 1975", Chemische Gesellschaft, 20, Nr. 1 (1973), 1-7.

Das Programm des Sozialismus wird verwirklicht. Zahlen-Fakten-Informationen. Aus Anlaß des VII. Parteitages der Sozialistischen Einheitspartei Deutschlands als Gemeinschaftsarbeit herausgegeben von der Staatlichen Zentralverwaltung für Statistik und dem Dietz Verlag (Berlin: Dietz Verlag, 1967).

Programme of the Socialist Unity Party of Germany. 9th Congress of the Socialist Unity Party of Germany, Berlin, May 1976 (Dresden: Verlag Zeit im Bild, 1976).

Protokoll der Verhandlungen des V. Parteitages der Sozialistischen Einheitspartei Deutschlands (Berlin: Dietz Verlag, 1959).

"Rahmenverträge - neue Form effektiver Zusammenarbeit von VVB und Hochschule", Das Hochschulwesen, Nr. 10 (1965), 647-655.

"Rechtsvorschriften zur Durchführung der in der Grundsatzregelung für die Gestaltung des ökonomischen Systems des Sozialismus in der Deutschen Demokratischen Republik im Zeitraum 1971 bis 1975 enthaltenen Aufgaben", Die Wirtschaft, Beilage 15 zur Nr. 19-20 (7. Mai 1970).

Die Redaktion, Die Wirtschaft, Nr. 14 (1968), 16.

"Reserven aufdecken und nutzen", Die Wirtschaft, Nr. 20 (1962), 15.

Rexin, M., "Die Entwicklung der Wissenschaftspolitik in der DDR", in Wissenschaft und Gesellschaft in der DDR (München: Carl Hanser Verlag, 1971), 78-121.

Richter, H.J., "Parameterpreisbildung unterstützt wissenschaftlich-technischen Fortschritt", Sozialistische Finanzwirtschaft, Nr. 17 (1975), 15-16.

"Richtlinien" see under "Gesetzblätter".

Rigo, E., "Kampf gegen Planrückstände und Schaffung des wissenschaftlich-technischen Verlaufs gehören zusammen", Die Wirtschaft, Nr. 23 (1970), 3.

Roesler, J., "Planung und Leitung des wissenschaftlich-technischen Fortschritts der DDR in der zweiten Hälfte der fünfziger Jahre", in Jahrbuch für Wirtschaftsgeschichte, Teil III 1970), 41-55.

Rompe, R., "Über die Zusammenarbeit zwischen Wissenschaft und Produktion", Einheit, Nr. 6 (1959), 718-724.

Rompe, R., "Probleme der Überführung wissenschaftlicher Forschungsergebnisse in die Produktion", Einheit, Nr. 6 (1962), 23-32.

Rothhaupt, F., S. Wittling, H. Jahn and H. Schenkel, Aufgaben und Organisation der technischen Vorbereitung der Produktion im sozialistischen Industriebetrieb (Berlin: Verlag Die Wirtschaft, 1962).

Rouscik, L., "Das neue ökonomische System der Planung und Leitung der Volkswirtschaft fördert die Wissenschaft und Technik", Die Technik, 18, Nr. 12 (1963), 765-768.

Rouscik, L., "Die optimale Betriebsgröße", Die Wirtschaft, Beilage zur Ausgabe Nr. 5 (1968), 4-10.

Rubenstein, A.H., "Der Einfluß organisatorischer Faktoren auf Entscheidungsprozesse in Forschung und Entwicklung - Der Fall dezentralisierter Großunternehmen", in J. Naumann (ed.), Forschungsökonomie und Forschungspolitik (Stuttgart: Ernst Klett Verlag, 1970), 345-363.

Rüger, A., "Die Bedeutung 'strukturbestimmender Aufgaben' für die Wirtschaftsplanung und -organisation der DDR", Sonderhefte des Deutschen Instituts für Wirtschaftsforschung, Nr. 85 (1969).

Rühle, W., "Die Einbeziehung der Grundlagenforschung in den volkswirtschaftlichen Reproduktionsprozeß", Wirtschaftswissenschaft, Nr. 3 (1968), 353-366.

Rumpf, W., "Staatshaushaltsplan 1963: Steigende Akkumulation - höhere Ausgaben für Produktion und Konsumtion", Deutsche Finanzwirtschaft, Nr. 24 (1962), 3-9.

Rytlewski, R., "Zu einigen Merkmalen der Beziehungen der DDR und der UdSSR auf dem Gebiet der Naturwissenschaft und Technik", Deutschland Archiv, Sonderheft (1973), 55-60.

Sabisch, H. and K. Hermsdorf, "Bedürfnis- und bedarfsgerechte Forschung und Entwicklung - ein Beitrag zur sozialistischen Intensivierung", Wissenschaftliche Zeitschrift der Technischen Universität Dresden, Nr. 1-2 (1976), 137-146.

Sack, K. and H. Pätzold, Die Planung der technischen Entwicklung im sozialistischen Industriebetrieb (Berlin: Verlag Die Wirtschaft, 1962).

Saizew, B. and B. Lapin, "Prognostizierung von Wissenschaft und Technik", Die Wirtschaft, Nr. 28 (1969), 23.

Schade, E., "Maßstab der Prämiierung muß die Planerfüllung sein", Die Wirtschaft, Nr. 33 (1967), 18.
Schauer, S., Ökonomisch begründete Preisrelationen (Berlin: Verlag Die Wirtschaft, 1965).
Schenk, F., Magie der Planwirtschaft (Köln and Berlin: Kiepenheuer und Witsch, 1960).
Scherer, F.M., "Firm Size, Market Structure, Opportunity, and the Output of Potential Inventions", American Economic Review, 55 (1965), 1097-1123.
Scherer, F.M., "Unternehmensgröße und technischer Fortschritt", Die Aussprache, Nr. 7 (1969), 169-174.
Scherzinger, A., Zur Planung, Organisation und Lenkung von Forschung und Entwicklung in der DDR (Berlin: Duncker und Humblot, 1977).
Schmidt, H. and R. Naumann, Wissenschaftliche Arbeitsorganisation (Berlin: Dietz Verlag, third edition 1973).
Schmidt and Reinhold, "Probleme der Planung der wissenschaftlich-technischen Entwicklung", Die Wirtschaft, Nr. 2 (1966), 17.
Schmiedeknecht, B., "Erfahrungen mit der Verteidigungsanordnung", Die Wirtschaft, Nr. 46 (1974), 13.
Schober, H.-H., "Die zweite Komponente der Planwirtschaft", Einheit, 2, Nr. 6 (1947), 540-544.
Scholz, N., "Regulierung über den Preis?", Die Wirtschaft, Nr. 50 (1973), 11.
Schulz, A., "Erfolge und Aufgaben der Zusammenarbeit mit Produktionsbetrieben", Das Hochschulwesen, Nr. 8-9 (1954), 31-39.
Schwarzbach, W., "Der Plan Wissenschaft und Technik muß stabil sein", Die Wirtschaft, Nr. 5 (1976), 21.
Schwertner, E., "Zur Intensivierung der wissenschaftlichen Arbeit an den Universitäten und Hochschulen", Das Hochschulwesen, Nr. 6 (1973), 162-164.
Schwertner, E., "Aktuelle Aspekte der Hochschulforschung", Das Hochschulwesen, Nr. 10 (1974), 296-299.
Schwertner, E. and A. Kempke, Zur Wissenschafts- und Hochschulpolitik der SED (Berlin: Dietz Verlag, 1967).
Science in the Service of the People (Berlin: Panorama DDR - Auslandspresseagentur, 1974).
Seickert, H., Produktivkraft Wissenschaft im Sozialismus (Berlin: Akademie Verlag, 1973).
Seickert, H., "Fortschritte und Hemmnisse der Überführung", Spektrum, Nr. 10 (1975), 3-4.
Seickert, H., J. Schindler and A. Selle, "Probleme der Überführung von Ergebnissen der Grundlagenforschung der Akademie der Wissenschaften der DDR in die materielle Produktion", Wirtschaftswissenschaft, Nr. 3 (1976), 419-427.
Seidel, D., Risiko in Produktion und Forschung als gesellschaftliches und strafrechtliches Problem (Berlin: Staatsverlag der Deutschen Demokratischen Republik, 1968).
Selbmann, F., "Maßnahmen zur Verbesserung der Arbeit auf dem Gebiet der naturwissenschaftlich-technischen Forschung

und Entwicklung und der Einführung der neuen Technik", Die Technik, Nr. 10 (1957), 660-668.

Sherwin, C.W. and R.S. Isenson, First Interim Report on Project Hindsight (Summary) (Washington, D.C.: Office of the Director of Defense Research and Engineering, 30 June 1966).

Sindermann, H., On the Directives for the Five-Year Plan for the GDR's National Economic Development 1976-1980, issued by the 9th Congress of the Socialist Unity Pary of Germany (Dresden: Verlag Zeit im Bild, 1976).

Slama, J. and H. Vogel, "Niveau und Entwicklung der Kilogrammpreise im Außenhandel der Sowjetunion mit Maschinen und Ausrüstungen", Jahrbuch der Wirtschaft Osteuropas, Heft 2 (1971), 443-483.

Slama, J. and H. Vogel, "On the Measurement of Technological Levels for the Soviet Economy", in Z.M. Fallenbuchl (ed.), Economic Development in the Soviet Union and Eastern Europe, Volume 1 (New York: Praeger, 1975), 197-220.

Slama, J. and H. Vogel, "Comparative Analysis of Research and Innovation Processes in East and West", in C.T. Saunders (ed.), Industrial Policies and Technology Transfer between East and West (Wien and New York: Springer, 1977), 103-120.

Slamecka, V., Science in East Germany (New York: Columbia University Press, 1963).

Sommer, R., G. Pöggel and G. Thürmann, Komplexe Rationalisierung und sozialistischer Wettbewerb (Berlin: Verlag Tribüne, 1966).

Spiller, H., "Universität und Wirtschaft", Das Hochschulwesen, 15, Nr. 12 (1967), 835-842.

Springer, A., "Die Effektivität der wissenschaftlichen Forschung erhöhen!", Einheit, Nr. 6 (1966), 818-820.

Staatliche Zentralverwaltung für Statistik (ed.), Bilanz unserer Erfolge. 20 Jahre DDR in Zahlen und Fakten (Berlin: Staatsverlag der Deutschen Demokratischen Republik, 1969).

Stalin, J., "Über die Aufgabe der Wirtschaftler", in Fragen des Leninismus (Berlin: Dietz Verlag, 1951), 390-401.

Stanke, V., Science and Society (Berlin: Panorama DDR - Auslandspresseagentur, 1976).

Statistical Pocket Book of the German Democratic Republic (Berlin: Staatsverlag der Deutschen Demokratischen Republik, various years).

Statistische Jahrbücher der Deutschen Demokratischen Republik (Berlin: Staatsverlag der Deutschen Demokratischen Republik, various years).

Statistisches Bundesamt Wiesbaden, Statistische Jahrbücher für die Bundesrepublik Deutschland (Stuttgart and Mainz: W. Kohlhammer, various years).

Steeger, H., "Der wissenschaftlich-technische Fortschritt in unseren sozialistischen Industriebetrieben und die Sicherung seines höchsten ökonomischen Nutzeffekts", Einheit, Nr. 10 (1962), 20-24.

Steeger, H. and G. Schilling, "Die Verbesserung des Systems der ökonomischen Stimulierung der neuen Technik - Ein Beitrag zur Beschleunigung des wissenschaftlich-technischen

Fortschritts in der Industrie der DDR", Wissenschaftliche Zeitschrift der Hochschule für Ökonomie Berlin, Nr. 2 (1961), 105-120.

Steenbeck, M., "Macht und Intelligenz im Wandel der Zeit", Einheit, Nr. 6 (1967), 674-685.

Stern, C., "Abbruch eines Denkmals", Zeit Magazin, Nr. 27 (Juni 1973), 2-7, 18.

Stoph, W., Die Durchführung der volkswirtschaftlichen Aufgaben. VII. Parteitag der Sozialistischen Einheitspartei Deutschlands, Berlin, 17. bis 22. April 1967 (Berlin: Dietz Verlag, 1967).

"Strategie der Großforschungszentren", Die Wirtschaft, Nr. 38 (1970), 6.

Strauss, C.-J., Der Meister. Material, Technologie, neue Technik (Berlin: Verlag Die Wirtschaft, 1961).

"Strukturpolitik", Die Wirtschaft, Nr. 27 (1968), 2.

Such, H., VVB und wissenschaftlich-technischer Fortschritt (Leipzig: Staatsverlag der Deutschen Demokratischen Republik, 1964).

Suchardin, S.V. (ed.), Wissenschaft als Produktivkraft (Berlin: VEB Deutscher Verlag der Wissenschaften, 1974).

Supranowitz, S., Kooperation in Forschung und Technik (Berlin: Staatsverlag der Deutschen Demokratischen Republik, 1968).

Sydow, W. (ed.), Ideen-Projekte-Produktionen. Aktuelle Fragen in Forschung und Entwicklung (Berlin: Verlag Die Wirtschaft, 1969).

Tannhäuser, S., "Konzentration des Forschungspotentials", Die Wirtschaft, Beilage zur Ausgabe Nr. 18 (1968), 16-17.

Taut, F., "Neue Erzeugnisse - Inhalt des wissenschaftlich-technischen Fortschritts", Die Wirtschaft, Nr. 7 (1969), 7.

Thießen, P.-A., "Probleme der Planung und Organisation der wissenschaftlich-technischen Forschung und Entwicklung in der Deutschen Demokratischen Republik", Einheit, Nr. 5 (1959), 592-606.

Traces. Technology in Retrospect and Critical Events in Science (Illinois Institute of Technology Research, Report Prepared for National Science Foundation, Washington, D.C., 1969).

"Überleitungsprozeß mit höchster Effektivität gestalten", Die Wirtschaft, Nr. 29 (16. Juli 1970), 2.

Uhlemann, G., "Trotz reduzierter Vorgaben noch zu viele Planteile", Die Wirtschaft, Nr. 48 (1966), 15.

Ulbricht, W., "Aufgaben der deutschen Wissenschaft bei der Schaffung der Grundlagen des Sozialismus", Tägliche Rundschau, Nr. 21 (25. Januar 1953), 5.

Ulbricht, W., Fragen der politischen Ökonomie der Deutschen Demokratischen Republik (Berlin: Dietz Verlag, 1955).

Ulbricht, W., Die Rolle der Deutschen Demokratischen Republik im Kampf um ein friedliches und glückliches Leben des deutschen Volkes (Berlin: Dietz Verlag, 1955).

Ulbricht, W., "Über die Arbeit der Sozialistischen Einheitspartei Deutschlands nach dem XX. Parteitag der KPdSU und die bisherige Durchführung der Beschlüsse der III.

Parteikonferenz", in Zur sozialistischen Entwicklung der Volkswirtschaft seit 1945 (Berlin: Dietz Verlag, 1960), 567-586.

Ulbricht, W., Das neue ökonomische System der Planung und Leitung der Volkswirtschaft in der Praxis (Berlin: Dietz Verlag, 1963).

Ulbricht, W., Das Programm des Sozialismus und die gesellschaftliche Aufgabe der Sozialistischen Einheitspartei Deutschlands, VI. Parteitag der Sozialistischen Einheitspartei Deutschlands, Berlin, 15. bis 21. Januar 1963 (Berlin: Dietz Verlag, 4. Auflage 1963).

Ulbricht, W., "Um die schnelle Steigerung der Arbeitsproduktivität auf der Grundlage höchster Leistungen von Wissenschaft und Technik", Neues Deutschland (16. Januar 1963), 4-5.

Ulbricht, W., Die Durchführung der ökonomischen Politik im Planjahr 1964 unter besonderer Berücksichtigung der chemischen Industrie. Referat auf der 5. Tagung des ZK der SED, 3. bis 7. Februar 1964 (Berlin: Dietz Verlag, 1964).

Ulbricht, W., Die nationale Mission der DDR und das geistige Schaffen in unserem Staat. Rede auf der 9. Tagung des ZK der SED, 26. bis 28. April 1965 (Berlin: Dietz Verlag, 1965).

Ulbricht, W., "Der Weg zur Durchführung der Beschlüsse des VII. Parteitages der SED auf dem Gebiet der Wirtschaft, Wissenschaft und Technik", Die Wirtschaft, Nr. 41 (1967), 3-9.

Ulbricht, W., "Die weitere Gestaltung des gesellschaftlichen Systems des Sozialismus", Die Wirtschaft, Beilage zur Ausgabe Nr. 44 (31. Oktober 1968), 10-35.

Ulbricht, W., Zum ökonomischen System des Sozialismus in der DDR (Berlin: Dietz Verlag, 1968).

Ulbricht, W., Die gesellschaftliche Entwicklung in der Deutschen Demokratischen Republik bis zur Vollendung des Sozialismus. VII. Parteitag der Sozialistischen Einheitspartei Deutschlands, Berlin, 17. bis 22. April 1967 (Berlin: Dietz Verlag, 1970).

Ulbricht, W., "Akademie und Großindustrie im Kampf um Pionierleistungen", Neues Deutschland (14. März 1970), 3.

Ulbricht, W., "Spitzenleistungen demonstrieren unsere ökonomische und politische Kraft", Die Wirtschaft, Beilage 9 zur Ausgabe Nr. 13 (26. März 1970), 14-16.

Ulbricht, W., "Für einen großen Aufschwung bei der Durchführung der Wissenschaftsorganisation in der chemischen Industrie", Die Wirtschaft, Beilage 11 zur Ausgabe Nr. 14 (1970), 9-16.

Ulbricht, W., "'Überholen ohne einzuholen' - ein wichtiger Grundsatz unserer Wissenschaftspolitik", Die Wirtschaft, Nr. 9 (1970), 8-9.

Usko, M., Hochschulen in der DDR (Berlin: Verlag Gebr. Holzapfel, 1974).

Verner, P., "Aus dem Bericht des Politbüros an die 14. Tagung des ZK der SED", Neues Deutschland (10. Dezember 1970), 4-9.

"Verordnungen" see under "Gesetzblätter".
"Verschwender von Volkseigentum am Pranger", Deutsche Finanzwirtschaft, Nr. 13 (1962), F9-F10.
"Verstärkt den Kampf um die Planerfüllung organisieren", Die Wirtschaft, Nr. 21 (14. Mai 1970), 2.
Wagner, U., Interessenkonflikte zwischen politischer Führung und Betriebsleitungen in sowjetischen Zentralverwaltungswirtschaften (Dissertation Marburg, 1967).
Wahl, D., "Arbeit im Kollektiv - Quelle der Produktivität", Neues Deutschland (19. April 1969), 10.
Waldenburger, M., Der Grundsatz "Neue Technik - neue Normen" (Berlin: Verlag Die Wirtschaft, 1967).
Walther, H., "Gewinnabschläge für nicht dem technischen Fortschritt entsprechende Erzeugnisse", Deutsche Finanzwirtschaft, Nr. 2 (1962), 3-5.
Wambutt, H., "Erfahrungen bei der sozialistischen Wissenschaftsorganisation in unserer chemischen Industrie", Einheit, Nr. 8 (1970), 1033-1045.
"Weitere Entwicklung der Naturwissenschaften", Nationalzeitung (13. März 1955), 2.
Wesselburg, F., "Gewinn und leistungsgerechte Entlohnung des Leiters", Wirtschaftswissenschaft, Nr. 10 (1965), 1585-1595.
Wiesner, R., "Prämienfonds unterstützt Strukturpolitik", Deutsche Finanzwirtschaft, Nr. 21 (1968), F1.
Wiesner, R., "Zu einigen Hauptaufgaben der Prämiierung", Die Wirtschaft, Nr. 39 (1968), 14-15.
Wikarski, S., "Wesen und Aufgabe der sozialistischen Großforschung", Einheit, Nr. 3 (1969), 380-385.
Wilczynski, J., Technology in Comecon (London: Macmillan, 1974).
"Wissenschaft und Technik brauchen langfristige Orientierung", Die Wirtschaft, Nr. 15 (1974), 2.
"Wissenschaft und Technik entscheiden wesentlich über das Wachstum der Volkswirtschaft", Statistische Praxis, Nr. 10 (1973), 498.
"Wissenschaft und Technik vor neuen Aufgaben", Die Wirtschaft, Nr. 5 (1969), 9.
"Das wissenschaftlich-technische Potential der DDR", Informatik, Nr. 3 (1973), 1.
Wittbrodt, H. and B. Gieltowsky, "Zur Entwicklung der Zusammenarbeit zwischen der Forschungsgemeinschaft und der Industrie im Jahre 1963", Spektrum, Nr. 5 (1964), 190-197.
Wöltge, H., "Für eine dauerhafte Verbindung zwischen Hochschule und sozialistischer Praxis", Die Wirtschaft, Nr. 16 (1968), 5.
Wolter, W., "Zur Ökonomie der Hochschulbildung", in A. Knauer, H. Maier and W. Wolter (eds.), Bildungsökonomie. Aufgaben-Probleme-Lösungen (Berlin: Verlag Die Wirtschaft, 1968), 133-172.
Wykowski, S., "Enge Verbindung von Grundlagenforschung und Überleitung der Forschungsergebnisse in die Praxis", Das Hochschulwesen, Nr. 4 (1976), 104-105.

Zahlenspiegel. Ein Vergleich Bundesrepublik Deutschland/Deutsche Demokratische Republik (Bonn: Bundesminister für innerdeutsche Beziehungen, 5. überarbeitete Auflage, Juni 1976).

Zander, R., "Vorschlag zu den Aufgaben eines Industriezweiginstitutes", Die Wirtschaft, Nr. 14 (1966), 6.

"ZIF des Maschinenbaus", Freiheit, Nr. 49 (26. Februar 1968), 3.

"Zur Entwicklung von Elektrotechnik, Elektronik und wissenschaftlichem Gerätebau", Statistische Praxis, Nr. 7 (1974), 4* (Beilage).

"Zur Lage der Chemiewirtschaft in der DDR", Wochenbericht, Nr. 49 (1976), 457-460.

Zur ökonomischen Theorie und Politik in der Übergangsperiode, 3. Sonderheft (Berlin: Verlag Die Wirtschaft, 1957).

GDR PUBLISHERS CITED

Akademie Verlag, Berlin
VEB Deutscher Verlag der Wissenschaften, Berlin
Dietz Verlag, Berlin
VEB Hermann Haack, Geographisch-Kartographische Anstalt
Panorama DDR - Auslandspresseagentur, Berlin
Staatsverlag der Deutschen Demokratischen Republik, Berlin
Verlag Tribüne, Berlin
Verlag Die Wirtschaft, Berlin
Verlag Zeit im Bild, Dresden
Volk und Wissen Volkseigener Verlag, Berlin

GDR PERIODICALS AND NEWSPAPERS

Die Arbeit
Arbeit und Arbeitsrecht
Chemische Gesellschaft
Chemische Technik
Deutsche Außenpolitik
Deutsche Finanzwirtschaft (later Sozialistische Finanzwirtschaft)
Deutsche Lehrerzeitung
Deutsche Zeitschrift für Philosophie
Dokumentation Panorama DDR
Effekt
Einheit
Energietechnik
Erfindungs- und Vorschlagswesen (later Der Neuerer)
Die Fachschule
Foreign Affairs Bulletin
Forum
Freiheit
GDR Export
Gesetzblatt der DDR
Der Handel
Das Hochschulwesen
Horizont

Informatik
Jahrbuch der DDR
Jahrbuch der Deutschen Akademie der Wissenschaften zu Berlin
Jahrbuch für Wirtschaftsgeschichte
Kraftfahrzeugtechnik
Marktforschung
Marktinformationen für Industrie und Außenhandel der DDR
Maschinenbautechnik
Mitteilungsblatt der Deutschen Akademie der Wissenschaften zu Berlin
Nationalzeitung
Neue Bergbautechnik
Neue Hütte
Neuer Weg
Der Neuerer (previously Erfindungs- und Vorschlagswesen)
Neues Deutschland
Plaste und Kautschuk
Probleme des Friedens und des Sozialismus
Sächsische Zeitung
Sozialistische Arbeitswissenschaft
Sozialistische Außenwirtschaft
Sozialistische Finanzwirtschaft (previously Deutsche Finanzwirtschaft)
Sozialistische Planwirtschaft
Spektrum
Staat und Recht
Statistische Praxis
Tägliche Rundschau
Die Technik
Technische Gemeinschaft
Vertragssystem
Die Wirtschaft
Wissenschaft und Fortschritt
Wissenschaft und Produktion
Wirtschaftsrecht
Wirtschaftswissenschaft
Wissenschaftliche Zeitschrift der Hochschule für Ökonomie, Berlin
Wissenschaftliche Zeitschrift der Humboldt-Universität zu Berlin
(1) Mathematisch-naturwissenschaftliche Reihe
(2) Ges.-Sprachwissenschaftliche Reihe
Wissenschaftliche Zeitschrift der Martin-Luther-Universität Halle-Wittenberg
Ges.-Sprachwissenschaftliche Reihe
Wissenschaftliche Zeitschrift der Technischen Hochschule Otto von Guericke, Magdeburg
Wissenschaftliche Zeitschrift der Technischen Universität Dresden
Wissenschaftliche Zeitschrift der Universität Rostock
Gesellsch.-sprachwissenschaftliche Reihe
Zeitschrift für Geschichte

Name Index

Subject Index